Handbook of ENTOMOLOGY

Handbook of ENTOMOLOGY

– Author –

Motiur Rehman Nomani

Wordpen Academics
Publishing Science That Matter

ISBN: 978-93-90438-62-4 (Hardback)

Published by : **WORDPEN ACADEMICS**
4736/23, Ansari Road, Darya Ganj,
New Delhi – 110 002
Phone: +91-011-35004336
E-mail: info@wordpenacademics.com

Printed at **:** **Replika Press Pvt. Ltd.**

Exclusive Distributor: ***Bio-Green Books, New Delhi***

Preface

Entomology is rooted in nearly all human cultures from prehistoric times, primarily in the context of agriculture (especially biological control and beekeeping), but scientific study began only as recently as the 16th century. The list of entomologists through recorded history is enormous, and includes such notable figures as Charles Darwin, Vladimir Nabokov, Karl von Frisch, and two-time Pulitzer Prize winner E. O. Wilson. Entomology is the scientific study of insects, a branch of arthropodology. At some 1.3 million described species, insects account for more than two-thirds of all known organisms, date back some 400 million years, and have many kinds of interactions with humans and other forms of life on earth. It is a speciality within the field of biology. Though technically incorrect, the definition is sometimes widened to include the study of terrestrial animals in other arthropod groups or other phyla, such as arachnids, myriapods, earthworms, and slugs. Like several of the other fields that are categorized within zoology, entomology is a taxon-based category; any form of scientific study in which there is a focus on insect related inquiries is, by definition, entomology. Entomology therefore includes a cross-section of topics as diverse as molecular genetics, behaviour, biomechanics, biochemistry, systematics, physiology, developmental biology, ecology, morphology, paleontology, anthropology, robotics, agriculture, nutrition, forensic science and more.

Today we are going to begin with the first point, the regulation of water balance, which may be insulted by substantial loss or gain of water. Insects are generally small creatures, which tends to put them in an unfavourable setting with respect to the ratio of body surface area to body volume. The idea is to recognize a relatively large surface across which water can be lost by transpiration. We will see that water loss across the body surface can not be entirely stopped; however, if it were not restricted to a level that can be balanced by water uptake, insects would not survive in the more arid environments. The first line of defence was the development of a cuticle relatively impermeable to water. The outer layer of an insect is the integument, which is made up of the epidermis and the cuticle. The epidermis is the outer cell layer. The epidermis lies on a basement membrante, and secretes the cuticle, which so far as I know is illustrated in virtually every book written on insect biology. There are three major sections of cuticle, endocuticle, exocuticle and epicuticle.

In standard drawings, we note the pore canal, cells of the epidermis and the basement membrane. The endocuticle provides the extensibility of the integument and combines toughness with flexibility. The rigidity of the hard parts of an insect, the appendages and so on, comes from the properties of the exocuticle. The relative impermeability of the cuticle to water is due to the presence of the epicuticle. The epicuticle is a complex structure that differs among various insects.

There is a cuticulin layer made of two components: a thin, dense outer layer and a thicker, still denser inner layer. Cuticulin is generally made of lipoprotein. A waxy layer lies atop the cuticulin. This is made of long-chain hydrocarbons, and esters of fatty acids and alcohols. There is an inner wax layer composed of strongly oriented long-chain molecules that generally form a monolayer only one molecule thick. The strong orientation provides the waterproof properties of the epicuticle. The outer wax layer is composed of wax molecules that are randomly orientated; on top of this is the cement layer. People think that this is a shellac-like substance that may serve to protect the wax. Terrestrial insects lose water by evaporation from the general body surface-a process called transpiration. The rate of water loss from the surface on any given insect is a function of both wind speed and also of the humidity of air. Transpiration is relatively independent of body temperature, up to some point. Water is lost from the body surface at a constant, but rather low, rate through a broad range of temperatures. At some temperature the cuticle reaches a temperature point called the transition point. This point varies from insect to insect, depending upon the chemical composition of the hydrocarbons and waxes in the epicuticle. Insects living in hot, arid habitats tend to have wax layers with higher transition temperatures than do insects living in cooler or more humid habitats. Some insects have two transitions points.

The uniqueness of this book hence lies in the editor's way of reconstructing the chapter under review by delving deep into the areas of the subject.

—Dr. Md. Motiur Rehman Nomani

Contents

1

Introduction to Entomology

Entomology

Entomology is the scientific study of insects, a branch of arthropodology. At some 1.3 million described species, insects account for more than two-thirds of all known organisms, date back some 400 million years, and have many kinds of interactions with humans and other forms of life on earth. It is a speciality within the field of biology. Though technically incorrect, the definition is sometimes widened to include the study of terrestrial animals in other arthropod groups or other phyla, such as arachnids, myriapods, earthworms, and slugs. Like several of the other fields that are categorized within zoology, entomology is a taxon-based category; any form of scientific study in which there is a focus on insect related inquiries is, by definition, entomology. Entomology therefore includes a cross-section of topics as diverse as molecular genetics, behaviour, biomechanics, biochemistry, systematics, physiology, developmental biology, ecology, morphology, paleontology, anthropology, robotics, agriculture, nutrition, forensic science and more.

History of Entomology

Entomology is rooted in nearly all human cultures from prehistoric times, primarily in the context of agriculture (especially biological control and beekeeping), but scientific study began only as recently as the 16th century. The list of entomologists through recorded history is enormous, and includes such notable figures as Charles Darwin, Vladimir Nabokov, Karl von Frisch, and two-time Pulitzer Prize winner E. O. Wilson.

Identification of Insects

Most insects can easily be recognized to order, such as Hymenoptera or Coleoptera. However, insects other than Lepidoptera are typically identifiable to genus or species only through the use of Identification keys and Monographs.

Because the class Insecta contains a very large number of species (over 330,000 species of beetles alone) and the characters separating them are unfamiliar, and often subtle (or invisible without a microscope), this is often very difficult even for a specialist.

Insect identification is an increasingly common hobby, with butterflies and dragonflies being the most popular.

Entomology for Beginners

Entomology is the study of insects. This page provides some basic information on insects. First a quick look at the anatomy of adult insects. Then a short story about the metamorphosis of insects.

Metamorphosis of Insects

In their development from egg to adult, insects change in form. The different instars are not alike, which is called metamorphosis. There are two general types of metamorphosis: simple metamorphosis and complete metamorphosis.

Simple Metamorphosis

In simple metamorphosis the wings develop externally during the larval stages. The larval stages, which are called nymphs, look very similar to the adult insect. There is no pupal stage.

Fig. : The figure shows the development from an egg via 5 larval instars to an adult bug.

Complete Metamorphosis

In a complete metamorphosis the wings develop internally during the larval stages. The larval stages look quite different from the adult. Between the last larval stage and the adult stage there is a pupal stage which usually is inactive.

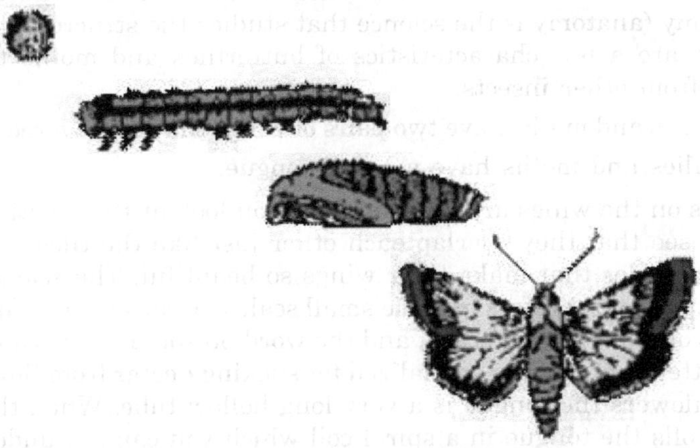

Fig. : The figure shows the development from an egg via the larval stage (caterpillar) and pupa to an adult butterfly or moth.

Butterflies for Beginners

Everybody loves butterflies. This page provides some basic information on butterflies. Actually, it is not only about butterflies but also about moths. We will learn how to recognize butterflies and we learn about their life cycle. There will be some information on how to rear butterflies, at home or in your classroom, and this page also has some proverbs and quotes about butterflies and some interesting butterfly facts.

What is a Butterfly?

Butterflies and moths are insects. There are many types of insects in the world and they have been divided in about 30 different groups which are called "orders". One of these orders are the "beetles", another order are the "flies and mosquitoes" and perhaps the most famous order are the "butterflies". The scientific name for this order is Lepidoptera and it includes both butterflies and moths.

Butterflies and moths are very similar. One difference is that butterflies are usually (but not always) very colourful and fly during daytime, while moths have usually (but not always) dull colours and fly at night. Another difference between butterflies and moths is based on the shape of heir antennae, and the way how their front and hind wings are connected.

If you want to know more about insects in general go to "entomology for beginners" (entomology is the name for the study of insects).

Butterfly Anatomy

If we want to know how butterflies differ from other insects we have to look at their anatomy (anatomy is the science that studies the structure of animals and plants). There are a few characteristics of butterflies and moths that set them clearly apart from other insects:

- Butterflies and moths have two pairs of wings that are covered in tiny scales.
- Butterflies and moths have a coiled tongue.

The scales on the wings are very small. If you look at them with a magnifying glass you can see that they overlap each other just like the tiles of a roof. It are these colourful scales that make their wings so beautiful. The scientific name for butterflies (Lepidoptera) refers to these small scales, because in old Greek language the word lepis or lepid means 'scale', and the word pteron means 'wing'. The tongue of an adult butterfly of moth is specialized for sucking nectar from flowers. To reach far into deep flowers the tongue is a very long hollow tube. When the butterfly is not eating it rolls the tongue in a spiral coil which you can see under the head of the butterfly.

Fig. : Side view of a butterfly. Clearly visible is the coiled tongue and the colourful wings. The drawing shows 3 legs but a butterfly has 6 legs just like any other insect.

Fig. : Top view of a butterfly. You can see that it has four wings (two pairs) and a pair of antennae. In butterflies the antennae are often knobbed.

To recognize a butterfly we have mentioned the coiled tongue and the colourful wings which are covered with scales. Apart from this the body of an adult butterfly or moth has the same body parts as in most other insects. The body has three major parts (head, thorax, and abdomen). The thorax has two pairs of wings (= 4 wings) and three pairs of legs (= 6 legs). The head has one pair of antennae (= 2 antennae).

Life Cycle of a Butterfly

When we talk about butterflies we are usually thinking about the colourful adult butterfly fluttering around in a garden or forest. But before reaching this beautiful adult stage it has to develop through a number of stages starting with a tiny egg. During its development from egg to adult, a butterfly passes several very different looking instars. This is called metamorphosis. The type of metamorphosis we find in butterflies is called "complete metamorphosis".

Complete Metamor Phosis

In a complete metamorphosis the wings develop internally during the larval stages. The larval stages of a butterfly is called "caterpillar" and it looks quite different from the adult butterfly. Caterpillars grow in a number of steps by shedding their skin (molting). Each instar is bigger than the previous one. The caterpillars are serious eaters and if there are a lot of caterpillars together they can do serious damage to a plant, eating all its leaves.

After the last larval stage there is a pupal stage. This pupal stage is inactive. Sometimes it can show some body movement (some wriggling) but it can not move around as it has no legs and often the pupa is fixed to the surface of a plant with some silk threads. Some pupa are "naked" but some species of butterflies have a pupa which is protected by a cocoon made of silk.

Inside the pupa the adult butterfly is developing and when it is ready the pupa opens and the butterfly emerges from it.

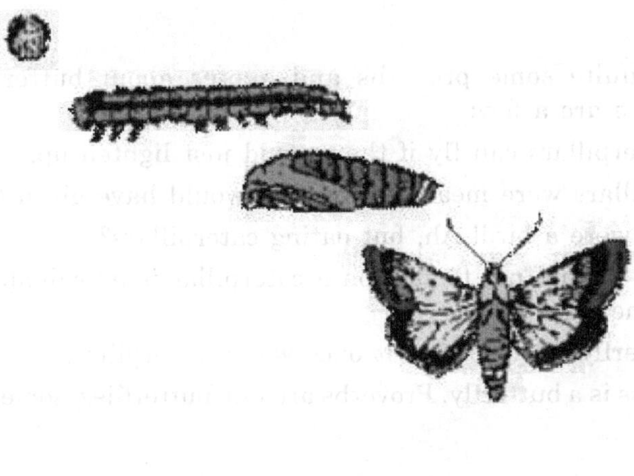

This figure shows the development from egg via caterpillar and pupa to an adult butterfly or moth.

Many people love the colourful adult butterfly but just as many don't like the caterpillars from which they develop. Caterpillars are often thought to be ugly and for farmers and gardeners the caterpillars can be a real pest destroying their crop or plants. But if you study them closely, even caterpillars have their beauty. Study them closely with a hand lens and see how they move and eat leaves.

Caterpillars

You have already learned that larvae of butterflies and moths are called caterpillars. You have also learned that insects have 6 legs. In an adult butterfly you can easily count these six legs, but in a caterpillar it looks like there are many more legs.

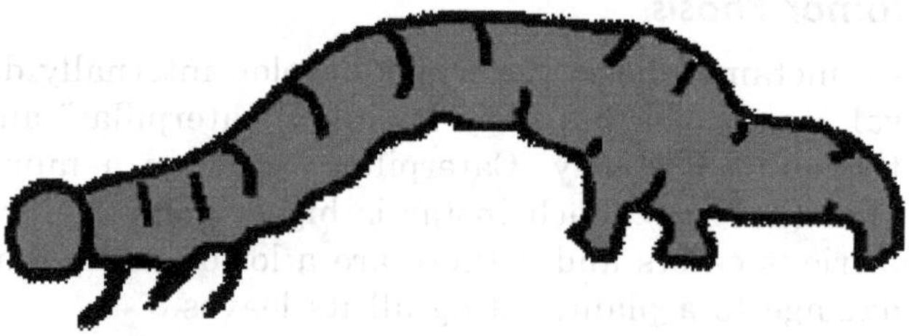

If you look a bit closer at a caterpillar (with a hand lens) you can see that it has two different types of legs. Just behind the head (on the thorax) it has 3 pairs of legs which are the real legs. These legs have joints and they have small claws at the end. The other legs which you see at the rearare not real legs. These are called the "prolegs" which are quite fleshy and have no visible joints. A caterpillar has 3 pairs of real legs on the thorax (just behind the head) and several pairs of fleshy 'prolegs' on the rear part of the body.

Proverbs

There are quite some proverbs and quotes about butterflies, moths and caterpillars. Here are a few:

- Even caterpillars can fly if they would just lighten up.
- If caterpillars were meant to fly, God would have given them wings.
- If only I were a bird! Ah, but eating caterpillars?
- Teaching a child not to step on a caterpillar is as valuable to the child as it is to the caterpillar.
- The butterfly often forgets it once was a caterpillar.
- Happiness is a butterfly. Proverbs are like butterflies, some are caught, some fly away

- The butterfly counts not months but moments, and has time enough.
- I dreamed I was a butterfly, flitting around in the sky; then I awoke. Now I wonder: Am I a man who dreamt of being a butterfly, or am I a butterfly dreaming that I am a man?
- You are like the butterfly that flies from flower to flower. Love is like a butterfly, hold it to tight it will crush, hold it to loose, it will fly.
- He dreads a moth who has been stung by a wasp.

Rearing Butter Flies

It is quite interesting and not very difficult to rear butterflies at home or in the classroom. You can start with any stage in the butterfly's life cycle, but often it is easiest to start with caterpillars because they are easier to find then the eggs. When collecting caterpillars you have to make sure that you know the host plant it is feeding on, because many caterpillars just eat one type of plant. if you try to rear them on other plants they will refuse to eat and will eventually die. To rear insects you need to prepare a living space for them from where they cannot escape and where you can easily observe them. You could use an empty aquarium in which you feed them with fresh leaves every day. Even better is to have a living plant in a pot to make sure that they always have fresh food. Keep the plant in a cage made from a wooden frame covered with mosquito netting so that the insects cannot escape.

Butterfly Facts

- In some countries certain caterpillars and pupa of butterflies are eaten. Usually they are deep fried and actually they taste quite nice.
- Silk is made from the cocoon of a moth, which called silkworm or silk moth. The scientific name of the silkworm is "*Bombyx mori*". The caterpillars of the silk moth feed on the mulberry tree.
- Adult butterflies usually have 4 wings, but there are also wingless species. There are moths where the females have no wings. Only the males can fly and have to find the female to mate.
- Caterpillars of the large blue butterfly live a part of their life inside an ant colony where they feed on the eggs and larvae of the ants. And the ants even protect the butterfly when it emerges from its pupa.
- The largest butterfly in the world has a wingspan of about 27 centimeters (11 inches).
- A biologist who studies butterflies and moths is called a "lepidopterist".
- A showy or frivolous person is called a "social butterfly".
- The "butterfly stroke" is a way of swimming which resembles a butterfly in flight.

Introduction to the Classification of Insects

Taxonomy is the study of the principles of scientific classification. In this page we will learn how insects are classified. First we will see where insects belong in the animal kingdom. Then we will find out how the different insects are sorted in groups.

Classification of Animals

The animal kingdom is divided in a number of groups called "phyla" (singular: phylum). Examples of phyla are:

- Protozoa (single-celled animals)
- Porifera (sponges)
- Nemathelminthes (roundworms)
- Mollusca (mollusks, snails, etc.)
- Arthropoda (crayfish, millipedes, centipedes, spiders and insects)
- Chordata, (fish, amphibians, reptiles, birds, and mammals).

Each phylum is subdivided in classes, for example the class Hexapoda (= insects). Classes are subdivided into orders, for example the order Coleoptera (= beetles). Orders are divided into families, families into genera (singular: genus), and genera are divided into species. Within the class Hexapoda there are over 750,000 different species of insects. The scientific name of a species is always a double name (the genus name, and a specific name). It should be written with a capital letter in the genus name and either in italics or underlined.

Example: Helicoverpa armigera or Helicoverpa armigera

An example of the classification of an insect:

Kingdom — Animal

Phylum — Arthropoda

Class — Hexapoda (= insects)

Order — Lepidoptera (= butterflies and moths)

Family — Noctuidae (= noctuids)

Genus — *Helicoverpa*

Species — *Helicoverpa armigera* (= American bollworm).

The Phylum Arthropoda

Some characteristics of the Arthropoda are:

- They have a so called exoskeleton. They do not have bones, but the hard outer covering supports the muscles.

- The appendages are jointed.
- The body is formed of a number of segments.

Characteristics of the Class Hexapoda (Insects)

Some characteristics of insects are:

Body:

- The body is divided into three distinct regions: head, thorax and abdomen

Head:

- One pair of antennae.
  * The antennae are usually used as tactile organs (= organs pertaining to the sense of touch) or as olfactory organs (= organs of smell).

Eyes:

  * Most insects possess one pair of compound eyes and sometimes some simple eyes called "ocelli".
- Mouthparts.
  * There is a big variety in types of mouthparts; biting, sucking, stinging, licking, etc.

Thorax:

- Three pairs of legs.
  * The thorax has three segments. These are called pro-thorax, meso-thorax and meta-thorax. Each segment has one pair of legs. The different parts of the leg are called coxa, trochanter, femur, tibia, and tarsus.

Note: some insects are legless, or have fewer than 6 legs. Some larvae have leg-like appendages on the abdomen.

- Often one or two pairs of wings.
  * The wings are borne by the second and/or third of the thoracic segments.

Note: Some insects are wingless.

Abdomen:

- The gonopore (genital opening) is at the posterior end of the abdomen.
- No appendages used for moving on the abdomen of adults (except in a few primitive insects).
- Sometimes there are some appendages at the end of the abdomen.

Classification of Hexapoda (Insects)

The class hexapoda is divided in two subclasses:

- Apterygota (= primitive wingless insects)

- Pterygota (= winged and secondarily wingless insects)

The subclass Pterygota is divided in two divisions:

- Exopterygota (= insects with a simple metamorphosis, without pupal stage)
- Endopterygota (= with a complete metamorphosis, including a pupal stage).

Metamorphosis

After hatching from the egg, an insect grows by a series of molts. After shedding the old skin they expand into a new larger one. This molting continues until the adult stage is reached. At each molt, some externally visible changes occur. This type of growing is called metamorphosis. The division of insects into apterygota, exopterygota and endopterygota is mainly based on differences in the type of metamorphosis.

The apterygota have no metamorphosis. Except for the size, all larval stages closely resemble the adults (which are wingless).

The exopterygota undergo a simple metamorphosis. In molting from egg, via the nymphal stages to an adult, there is a gradual change in the external appearance. The late nymphal stages already show the development of wing pads. But only in the last molt functional wings are developed. The nymphs usually have the same feeding habits as the adults. In the endopterygota there is a complete metamorphosis. In these insects the external changes during the life history are the greatest. The eggs hatch into larvae which feed actively during the different instars. The larvae may or may not have legs.

The development of wings is not visible during the larval stages. After several molts a pupa is formed. A pupa is an inactive stage, it does not feed and it does not move. Sometimes the pupa is protected by a cocoon of silk, or it is found in an earthen cell in the soil. During this pupal stage big changes take place internally. After the pupal stage, a highly active winged adult appears. Often, the larvae and the adults live in different types of habitat and use different types of food.

Orders of Insects

Orders marked with a (*) are important because they contain some agricultural pests.

Apterygota

Order Thysanura	Bristletails
Order Diplura	Diplurans (Two-pronged Bristletails)
Order Protura	Proturans
Order Collembola	Springtails

Exopterygota

Order Ephemeroptera	Mayflies
Order Odonata	Dragonflies and Damselflies
Order Orthoptera *	Grasshoppers, Locusts and Crickets
Order Dictyoptera	Cockroaches and Mantids
Order Grylloblattodea	Rock crawlers
Order Phasmida	Stick insects and Leaf insects
Order Dermaptera	Earwigs
Order Isoptera *	Termites
Order Embioptera	Web-spinners
Order Plecoptera	Stoneflies
Order Zoraptera	Zorapterans
Order Psocoptera	Psocopterans (Psocids, Booklice)
Order Mallophaga	Chewing lice (Biting lice)
Order Anoplura (= Siphunculata)	Sucking lice
Order Thysanoptera *	Thrips
Order Hemiptera	
suborder Heteroptera *	Bugs
suborder Homoptera *	Cicadas, Hoppers, Psyllids, Whiteflies, Aphids, and Scale insects

Endopterygota

Order Neuroptera	Alderflies, Dobsonflies, Fishflies, Snakeflies, Lacewings, Antlions, and Owlflies
Order Coleoptera *	Beetles
Order Strepsiptera	Twisted-winged parasites (Stylopids)
Order Mecoptera	Scorpionflies
Order Trichoptera	Caddisflies
Order Lepidoptera *	Butterflies and Moths
Order Diptera *	True Flies
Order Siphonaptera	Fleas
Order Hymenoptera *	Sawflies, Ichneumons, Chalcids, Ants, Wasps, and Bees

Identifying Insects

When trying to identify an unknown insect you will always first try to determine its correct Order. This can be done with the help of a key. You will need a good hand lens to observe some of the smaller parts of the insect, for example to count the number of segments in the tarsi, or to have a close look at the antennae. Click here to learn more about using a key to identify insect orders.

Good, Bad or Neutral?

We have just seen how insects can be classified in different orders. But there are other ways of grouping insects for example from the farmers' point of view. Farmers will usually classify insects in 3 groups, depending on their behaviour in the farm:

- Pests
- Beneficial insects
- Neutral insects.

Pest

Whether an insect species is a pest depends on the situation. A definition of "pest" is: animals causing damage or annoyance to man, his animals, crops or possessions, such as insects, mites, nematodes, rodents, birds. This means that a certain insect could be a pest in one situation, but the same insect would be neutral in another situation.

For example the caterpillars of Diamondback moth (*Plutella xylostella*) feed on cabbage and other plants of the Cruciferae family. A farmer who grows cauliflower or kale will therefore consider it a pest. But for a farmer who grows potatoes or bananas the Diamondback moth is an innocent, neutral insect.

Beneficial Insects

Some insects are beneficial to the farmer, because they are the natural enemies of other insects. Predatory insects feed on other insects and in this way they help control pest insects. For example the Assassin bug kills caterpillars and Ladybird beetles feed on aphids. Other insects are beneficial because they help with the pollination of plants, e.g. bees and bumble bees. There are insects that produce useful products, for example honey or silk. And in many countries insects are used as food.

Neutral Insects

If an insect is not a pest and not beneficial than we can call it neutral. But again, it really depends on the context. In a rice field a mosquito can be considered a neutral insect (it doesn't harm the crop), but in your bedroom you will call it a pest.

The History of Forensic Entomology

In an experiment famous as much for its demonstration of scientific method as for its contribution to entomology, Francesco L. Redi (1668) studied rotting meat that was either exposed to or protected from flies. From his analysis of subsequent blow fly infestation, he refuted the hypothesis of the "spontaneous generation" of life. Up to that time, it was generally believed that under the right conditions maggots came from rotten meat. Later, Bergeret (1855), near Paris, France, was the first westerner to use insects as forensic indicators. The body of a baby was found behind the plaster mantle in a house, and an investigation was begun. Bergeret determined that the assemblage of insects associated with the corpse pointed to a state of decay that dated back several years; consequently, the question of guilt was thrown upon the earlier occupants of the house, and not upon the current ones.

Bergeret's methods and materials were quite similar to one of the main medicocriminal entomological techniques still in use today; that is, the successive colonization of a corpse by a predictable succession of arthropod species. Between 1883 and 1898, J. P. Megnin in France published a series of articles dealing with medicocriminal entomology. The most famous of these, *La Faune des Cadares*, served in large part to make the medical and legal professions aware that entomological data could prove useful in forensic investigations.

Although entomologists are most familiar with the references cited above, medicocriminal matters in the Far East predate these considerably. In 1235 A.D., Sung Tz'u, a Chinese "death investigator," wrote a book entitled *The Washing Away of Wrongs* in which forensic science as known at that time was detailed. In this text, what was probably the first actual medicocriminal entomology case was recounted. A murder by slashing occurred in a Chinese village, and the local death investigator was deputized to solve the crime. After some fruitless questioning, the investigator had all villagers bring their sickles to one spot and lay them out before the crowd.

Flies were attracted to one of the sickles, probably because of invisible remnants of tissue still adhering to it, and the owner subsequently broke down and confessed to the crime. In other portions of the text, Sung Tz'u demonstrated knowledge of blow fly activity on bodies relative to those orifices infested, the time of such infestation, and the effect of trauma on attractiveness of tissue to such insects.

Any analytical system is as reliable as is the data upon which it is founded, and forensic entomology is no exception. Because accurate identification of necrophilous arthropods is of paramount importance, few repeatable results could be obtained before adequate taxonomic work had been accomplished on the invertebrates (the insects and related animals) in question. Taxonomy and systematics comprise the science describing, classifying, and proposing evolutionary relationships of the various forms of life.

Although many synanthropic (strongly associated with human activity) flies (such as Drosophila, Musca, Muscina, Ophyra, Stomoxys, and others) are not encountered frequently in typical forensic investigations, other species assume great importance. Carrion (dead tissue) feeding blow flies (Calliphoridae) and flesh flies (Sarcophagidae) are those most useful in death investigations. Aldrich's (1916) monograph on the Sarcophagidae made use of distinctive male genitalia, thereby enabling entomologists to identify adult male specimens from this important family.

This concept involved the so called "lock-and-key" arrangement in many insects that facilitates reproductive isolation between species. The male copulatory organs of each kind (species) of higher flies are composed of unique, complex structures that are used as key characters to enable specific determination. This adaptation has been applied with equal success to the forensically important blow flies.

Twenty years later, Knipling (1936) published descriptions and keys to many common early (first instar) maggots of flesh flies. Although considerable work had been done on the blow fly fauna of North America (for instance, Knipling 1939), Hall's 1948 monograph, *The Blowflies of North America,* made possible the accurate identification of adults and mature larvae of most species of this family as well.

Although very few new (that is, previously unrecognized) North American calliphorid species have been described recently, efforts have been devoted to accumulate improved distributional information (Hall and Townsend 1977, Hall 1979, Goddard and Lago 1983). More research is needed on accurate identification of the critical larval and pupal stages (those most frequently collected in death investigations). At present, first instar blow fly larvae (the stage that hatches directly from the egg) generally are not identifiable to species, and second instars (the next maggot stage) can be identified accurately only on occasion.

The situation is somewhat better with respect to third instar or prepupal larvae (the largest maggot stage, and that most commonly observed), but only if such specimens are preserved properly. Even so, a significant number of indigenous blow flies cannot be identified at present as immatures. This is currently an area of active research, and to this end the relatively new technique of scanning electron microscopy is being applied.

Because of the medicocriminal requirement for reliable data on rates of larval development, considerable effort has been expended to measure such intervals. Anecdotal information on blow flies contained in earlier works was largely supplanted by Hall's (1948) rearing data, and the latter has been refined for some forensically important species to degree hour status. Because insects are coldblooded animals, their rate of development is more or less dependent on ambient temperature.

Research has shown that for each species there generally is a threshold temperature below which no development takes place.

As temperature rises above this threshold, a certain amount of time is required for the insect to attain defined stages of development (for instance, from the newlylaid egg through the second instar maggot). Because this heat is accumulated as "thermal units," it can be calibrated and described as "degreedays" or "degreehours," depending on the accuracy of temperature readings and time period involved.

However, most laboratory rearings (upon which the degreehour data are developed) have been done at constant temperature, so additional research will be necessary to establish correlations between these data, typical fluctuating field temperatures (warmer during the day and cooler at night), and the average daily measurements frequently reported from weather stations. Retrospective weather records from the nearest weather recording station (such as an airport) are those most often used in medicocriminal evaluations.

Access to the scientific literature pertaining directly to medicocriminal entomology has been facilitated by two recent bibliographies. An initial guide to entomological involvement in forensic pathology, plus a selected bibliography, was provided by Meek et al. (1983). A bibliography of all publications dealing wholly or in large part with medicocriminal entomology worldwide was compiled by Vincent et al. (1985). The latter paper contained 329 references and was current through 1983; therefore, the actual body of literature pertaining to this subdiscipline of forensic entomology is not large when compared to many other biological or legal subjects. The first textbook devoted to forensic entomology was published in 1986: *A Manual of Forensic Entomology*. This is an excellent reference for the entomologist, and it brings together in one place all the salient information contained in the literature on this subject.

A procedural guide, *Entomology and Death*, was published in 1990 and is intended for crime scene investigators and other forensic specialists.

Scale Insects

Scale insects are a diverse group of insects in the order Hemiptera. There are about 6,000 species of scale insects in 21 families worldwide. About 1,000 species occur in North America. The three most common families of scale insects are the armoured scale, the soft scale, and the mealybugs. Most of the pest species belong to one of these three families.

Armoured Scale Insects: Armoured scales are the smallest of scale insects, ranging in size from 1 to 3 millimetres. The body of the scale insect is protected by a cover (the armour) made from wax secreted by the insect and cast skins (exuviae) of previous growth stages. One must remove the hardened wax cover to expose the body of the insect. The exposed body usually is yellow or orange, but may have a pink or red colour to it. This cover also protects the eggs laid by the female. Armoured scale insect covers vary from circular to elongate or oyster shell-shaped. Male and female covers may differ in size and shape for the same species. The cover of the female is generally largest. Boisduval scale and fern scale are common armoured scale insects attacking flowers and foliage plants.

Most armoured scale insects reproduce sexually. The eggs hatch beneath the protective scale cover and the first instars, commonly called "crawlers", migrate to the new growth to settle and feed. Armoured scale females lose their legs at the first molt and are sessile for the rest of their lives. Females develop through three instars and males develop through five. Armoured scales may overwinter as eggs, nymphs, or adult females. Adult males are usually present about two weeks in each generation. Some armoured scales have four generations per year.

Soft Scale Insects: Soft scales differ from armoured scales in that they do not secrete a waxy covering that is separate from the body. If wax is present, it adheres tightly to the body of the female and cannot be easily separated from it. Most soft scales produce a thin, glassy wax that does not obscure the colour or form of the female soft scale.

Soft scales are fairly large (2 to 6 millimetres long) and can be distinguished by their larger size, round or oval body outline, and convex or hemispherical profile. Soft scale females vary from flat to almost spherical. Often different host plants will alter the body form of a single species so much that taxonomists have described the different forms as separate species. If one turns the adult soft scale over, legs, antennae and thread-like mouthparts are readily visible with the aid of a microscope. Three common soft scales found in greenhouses and interior plantscapes are the brown soft scale, hemispherical scale, and tessellated scale.

Soft scales may reproduce sexually or parthenogenically and every female may be capable of producing progeny without fertilization. Tremendous populations can develop during a single growing season. Most outdoor species have one generation per year. Females either lay eggs or give live birth, depending on the species. There are three instars in the females and five instars in the males. In warmer climates and in greenhouses, species with multiple generations may have all stages present simultaneously throughout the year.

Key to Scale Insects

1. Body of insect protected with a wax or vanish-like covering that is not attached to the body of the insect; the insect can be seen by lifting off this covering. Armoured scale insects.
2. Fern Scale. Scale covering of female brownish and oystershell shaped; male cover snow-white with three ridges down the back.

3. Hemispherical scale. Mature individuals noticeably convex or hemispherical in side view, shiny brown in colour; sometimes hundreds of eggs found under the body.
4. Brown soft scale. Body yellowish-green to yellowish-brown in colour; often mottled with brown spots.

Boisduval Scale

Scientific Name: *Diaspis boisduvali* Signoret

Class: Insecta

Order: Hemiptera

Family: Diaspididae

Description

Adult Females: Cover of the adult female boisduval scale is circular or oval, thin, flat, white to light yellow, semitransparent, and 1.2 to 2.25 millimetres in diameter. Exuviae central to subcentral in position.

Other Stages: The male cover is somewhat rectangular in shape, elongate, about 1 millimetre in length, and snow white with three ridges down the back.

Eggs: Boisduval scale eggs are oval and pale yellow to pale orange.

Nymphs: Newly settled nymphs are pale orange.

Biology

Host Plants: Cattleya and Cymbidium orchids are the most common hosts, but this scale insect thrives on the leaves and bark of other orchids, palms, banana, and cacti.

Damage: New growth is frequently attacked prior to opening, and leaves may be killed if the infestation is dense. Lighter infestations produce irregular chlorotic spots that mark feeding sites and spoil the appearance of the leaves.

Life Cycle: Under greenhouse conditions, boisduval scale can complete its life cycle in about 50 days. Females may produce up to 200 eggs in a lifetime.

Eggs hatch in 5 to 7 days and most of the crawlers settle to feed within 24 hours after hatching.

The first stage lasts about 10 days, the second about 8 to 10 days for females and 15 for males. After mating, females retain their eggs for a short period, then oviposit. Because females produce eggs over a long period, generations overlap, and usually all stages can be found at any one time. Crawlers can settle anywhere on the leaf, but seem to prefer the midrib and major veins.

Management Strategies

Try to purchase plants from a supplier who does not have a scale infestation.

Pesticides: When scales are encountered, horticultural oils give adequate control without excessive phytotoxicity. Two thorough treatments 2 weeks apart should give good control. Boisduval scale crawlers often establish themselves near the base of the plant and in leaf sheaths making control difficult. Also, both upper and lower leaf surfaces are attacked, so thorough coverage of any treatment is essential. For specific control recommendations, consult the Cooperative Extension Service.

Fern Scale

Scientific Name: *Pinnaspis aspidistrae* (Signoret)

Class: Insecta

Order: Hemiptera

Family: Diaspididae

Description

Adults: Female fern scale armour is oystershell or pear shaped, flat, light brown with the crawler cast skin a paler brown. Sometimes the second stage armour is also paler than the adult armour.

They are 1.5 to 2.5 millimetres long. The male armour is white felted, three-ridged, and the crawler cast skin is beige to yellowish brown. Adult males are tiny, two-winged, gnat-like insects that are easily overlooked.

Eggs: Eggs of armoured scales are usually oval and about 0.2 millimetre long. They are laid in groups under the female armour.

Crawlers: Fern scale crawlers are about 0.2 millimetre long, flat, and yellow with red eyes. The legs and antennae are well developed.

Nymphs: Female second-stage nymphs secrete an oval, pale brown armour about 0.8 millimetre long. Male second stage nymphs secrete a white armour which has three long ridges. Mature male second stage nymphs are about 1 millimetre long.

Biology

Host Plants: Fern scales mainly infest true ferns (not asparagus ferns) and liriope. This pest has been recorded from numerous foliage plants, citrus, and other woody ornamental trees and shrubs in Florida.

Damage: Infested ferns are disfigured by the presence of male second stage armour which is conspicuous against the dark green foliage. Ferns in commercial production sometimes tolerate a tremendous scale population with little noticeable reduction in vigour or colour. Feeding by female scales causes yellow spots on some fern varieties and on liriope. Males do not feed beyond the second stage of development.

Life Cycle: Little specific information is known about the biology of fern scales. Female scales lay their eggs under the armour. The female dies after the last egg is laid. Tiny crawlers hatch from the eggs and eventually emerge from under the mother's armour. The crawlers move about until they begin to feed by inserting their long, thread-like mouthparts into the leaf and sucking out nutrients. The insect molts into a second stage which begins to secrete a waxy material from under the rear of the first stage (crawler) cast skin. Eventually these insects molt into the adult stage. Female scales begin to secrete the adult armour at the rear of the second stage armour. Males emerge from their second stage armour as tiny, gnat-like insects that crawl or fly to female scales to mate. The armour remains fastened to the plant long after the scale insect leaves (male) or dies inside (female). When populations become dense, females tend to lay male eggs so that heavily infested plants become conspicuously spotted by second stage male armour.

Management Strategies

Try to purchase plants from a supplier who does not have a fern scale infestation.

Pesticides: When fern scales are encountered, horticultural oils give adequate control without excessive phytotoxicity to ferns. Two thorough treatments 2 weeks apart should give good control. Ferns are notoriously sensitive to pesticides. Whenever treating ferns and other sensitive plants, treat at a time that the pesticide will be dry on the foliage before the plants are exposed to full sunlight. For specific chemical control recommendations, consult the Cooperative Extension Service.

Hemispherical Scale

Scientific Name: *Saissetia coffeae* (Walker)

Class: Insecta

Order: Hemiptera

Family: Coccidae

Description

Adults: Depending on the host plant, the adult scale may vary in size. The scale varies from 4.5 millimetres on Cycas to as small as 2.0 millimetres on Asparagus fern. Relatively hemispherical, brown, smooth, and shiny, the scale may resemble a miniature army helmet. Young females may have a pattern of ridges in the form of the letter "H" on the dorsal surface.

Eggs: The oblong, pinkish beige eggs are about 0.7 millimetre long and are protected by the mother's body in a mass of hundreds.

Crawlers: The flat, pinkish beige crawler is about 1.0 millimetre long with two red eye spots. The antennae and legs are short and spindly. There are two setae protruding from the rear.

Nymphs: Nymphs are semitransparent, light yellow or pink, and flat (young) to humped (older). They are so closely attached to the host plant that the legs and antennae are concealed. Two pale lines start at the margin on each side and fade in intensity toward the middle. Nymphs and young adults may exhibit the characteristic "H" pattern of the black scale group.

Biology

Host Plants: The hemispherical scale insect has a wide range of hosts including bamboo, camellia, chrysanthemum, crape myrtle, croton, ferns, figs, gardenia, honeysuckle, lily, orchids, and zamia.

Damage: The secretion of honeydew and subsequent formation of sooty mold detracts from the beauty and commercial value of the plant. Feeding causes the plant to become stunted and lose its leaves.

Life Cycle: The hemispherical scale insect is a tropical insect that has become a common greenhouse pest. In warmer temperate areas, the scale can live outside. Unlike many other scales, the hemispherical scale secretes little wax. However, the scales do produce much honeydew which allows sooty mold to grow. Males are unknown and reproduction is by parthenogenesis. Each female may deposit up to 1000 eggs. After laying the eggs, the female dies and her body shrinks to form a cup or helmet over the eggs.

Development for the egg stage and the three nymphal instars requires at least 40 days and may take as long as 105 days. There may be several generations each year in the greenhouse, with all stages being present at any one time. The nymphs do not move much once they have begun to feed. The hemispherical scale insect has a low rate of parasitism indoors, and it often becomes a serious pest.

Management Strategies

Biological Control: The most important parasite of hemispherical scale in the United States is *Metaphycus helvolus* Compere.

Pesticides: Hemispherical scale insects are relatively difficult to control because the eggs and young nymphs are protected by the body of the mother. For specific chemical control recommendations, consult the Cooperative Extension Service.

Brown Soft Scale

Scientific Name: Coccus hesperidium Linnaeus

Class: Insecta

Order: Hemiptera

Family: Coccidae

Description

Adults: Living adult female brown soft scales are pale yellowish green to yellowish brown, often mottled with brown spots. Older females are brown. The body is usually oval in outline, 2.5 to 4 millimetres in length and slightly convex in profile. The shape tends to vary according to position on the host plant. Males look like tiny two-winged wasps or flies and are rare.

Other Stages: Crawlers and young nymphs are yellow and almost flat in profile. Parasitized nymphs are dark brown to black and convex. Crawlers have well-developed legs and antennae and are quite active until settling. Older nymphs are sessile.

Biology

Host Plants: Brown soft scale has been reported feeding on hundreds of different plants. It can survive on most greenhouse plants, but seems to prefer perennials over annuals. Ferns are a favoured host.

Damage: Brown soft scale is probably the most frequently encountered scale on plants indoors. Infestations of brown soft scale can become so heavy as to encrust the stems and petioles of their host plant. They also settle on leaves, usually along midribs and occasionally on the fruit. Large colonies remove large quantities of plant fluids and can cause wilting, but they seldom kill their host. Immatures and adults produce much honey dew that serves as a medium for the growth of sooty molds. These fungi inhibit photosynthesis and make infested plants unsightly. Other insects such as ants and wasps are attracted to feed on the honeydew.

Life Cycle: Females are ovoviviparous and retain the eggs in the body until hatching. Brown soft scales are born as active crawlers but remain under the body of the female for a short time before emerging and selecting a feeding site to settle and complete their development. Females molt twice before reaching maturity. Males undergo four molts before emerging as winged adults, but are rarely seen. All stages are found throughout the year in warmer regions and in greenhouses. Brown soft scales can complete three to seven generations per year depending on temperature. It takes about 60 days to complete a generation.

Management Strategies

Brown soft scales are surprisingly difficult to control even though there is an external egg stage and only a few very young nymphs are protected by the body of the mother.

Biological Control: Although reported as a pest species of many host plants in many countries, outdoors it may be suppressed by natural enemies in many areas. *Scutellista cyanea* Motschulsky is a common parasite and *Metaphycus luteolus* Timberlake controls brown soft scale in California. The lady beetle *Ryzobius lophanthae* is a soft scale predator.

Pesticides: For specific chemical control recommendations, consult the Cooperative Extension Service.

Tessellated Scale

Scientific Name: *Eucalymnatus tessellatus* (Signoret)

Class: Insecta

Order: Hemiptera

Family: Coccidae

Description

Adult Females: Living adult females are flattened, reddish brown to dark brown, and 4-5 millimetres in length. The body is usually somewhat oval in outline, with the anterior end angulate and the posterior broadly rounded, but the shape may change due to position on the plant. When viewed under a hand lens or magnifying glass, the dorsal surface of the body has a checkered or cellular pattern, divided by a smooth longitudinal median ridge.

Crawlers: Tessellated scale crawlers are reddish brown, flat, elongate oval, and widest in the middle area of the body.

Nymphs: Settled nymphs are light golden in colour, flat, oval, and adhering so closely to the leaf surfaces as to be nearly transparent. Males are unknown.

Biology

Host Plants: Tessellated scale occurs most commonly on palms and several species of *Ficus*, but it has been found on a wide variety of host plants.

Damage: Tessellated scale is primarily a leaf-infesting soft scale and a greenhouse pest. Heavy infestations weaken the plant but seldom kill it. Infestations are often of economic importance in commercial nurseries whenever pest management programs are neglected.

Life Cycle: One or two generations per year, and in greenhouses generations will overlap. Tessellated scale is parthenogenetic and ovoviviparous, giving birth to living young. Males are unknown.

Management Strategies

Biological Control: The parasitic wasp *Metaphycus stanleyi* Compere is a natural enemy of tessellated scale.

Pesticides: As with other scale insects, tessellated scale is best controlled in the immature stage. For specific chemical control recommendations, consult the Cooperative Extension Service.

Aphids

Aphids are small soft-bodied, generally sluggish insects that have piercing-sucking mouthparts, which are inserted into the phloem tissue and remove fluids. Aphids are the only insects that have "honey tubes," or cornicles, on the abdomen. The cornicles, which project beyond the body of aphids can make them appear jet-propelled. Cornicle shapes and colours are used by taxonomists to help separate aphid species.

The most common aphids on greenhouse crops are green peach aphids (*Myzus persicae*), and melon/cotton aphids (*Aphis gossypii*). Both melon aphids and green peach aphids can feed on very large numbers of host plants. Both species are generally green insects, but the colour often varies. Melon aphids can vary from light green to very dark green, and green peach aphids from light green to nearly pink.

In greenhouses and tropical areas, all aphids are usually females that produce live young. The immature aphids are called nymphs. Each female can produce about 50 to 200 nymphs during her life span, and these nymphs (which are all females) can begin reproducing in 7 to 10 days. Adult aphids may or may not have wings, depending on population density and/or host plant condition. The winged forms are able to disperse within the greenhouse, and are also able to fly into the greenhouse from outdoor crops, overwintering sites, etc. Most of this flight is not directed. That is, aphids fly in the direction of the wind.

Plants must be inspected frequently to find localized aphid infestations. Yellow sticky traps will capture some flying winged aphids, but most aphids do not have wings.

Look for the insects themselves, which may be visible on plant stems or partially hidden on leaf undersides. Be alert as plants mature, because buds and flowers are favoured areas for aphid feeding. Small white skins left on plants by aphids during the molting (growth) process are other signs of an aphid infestation. Aphids also produce sticky liquid called honeydew, containing sugars and other chemicals, which can cover leaves and flowers. Whiteflies, mealybugs and soft scales also produce honeydew.

In addition to direct injury through feeding and being a nuisance, aphids are able to transmit several viruses, which cause diseases of both vegetable and ornamental plants. To date, however, virus problems caused by aphids on greenhouse ornamentals seem to be minimal.

Most growers control aphids using conventional or biorational pesticides. However, pesticide-resistant aphid populations are becoming common, and growers are sometimes having great difficulty in controlling them with insecticides. Aphids on buds or in flowers are especially difficult to deal with, because even after being killed the dead aphids remain.

There are several biological control options, including parasites and predators. Despite the widespread availability of biological control agents from commercial sources, biological controls are not widely used on ornamental plants. Most biological controls in greenhouses are used on vegetables.

Chrysanthemum Aphid

Scientific Name: Macrosiphoniella sanborni

Class: Insecta

Order: Hemiptera

Family: Aphididae

Description

Adults: The winged adults are about 2 to 2.5 millimetres long; soft bodied; and dark, shining mahogany brown. Due to their dark colour they were once called "blackflies." Wingless adults are only 1.5 millimetres long. Small, black, sturdy cornicles are found on the end of the body.

Nymphs: Nymphs resemble smaller versions of the adult aphids (0.6 to 1 millimetres). Nymphs have a dull, brick-red bodies with relatively long legs and antennae. The cornicles are short and dark. The outer two-thirds of the legs and antennae are gray (young nymphs) to dark gray. Older nymphs have proportionally longer cornicles. Those destined to be winged adults have wing buds in the later instars.

Biology

Host Plants: Chrysanthemum is the only known host for this aphid in North America, but it has been reported on a few related plants in Asia.

Damage: The chrysanthemum aphid feeds by piercing the plant surface with its threadlike mouthparts to suck out plant juices. They gather about the terminal buds and feed on the new growth.

This feeding causes distorted growth and the leaves may be covered by the feeding aphids' honeydew and cast skins. Sooty mold may grow on the honeydew giving the leaves and stems a black appearance. Chrysanthemum aphids are able to transmit chrysanthemum vein mottle virus and chrysanthemum virus B.

Life Cycle: Only female chrysanthemum aphids are known. They reproduce by giving live birth to more females, without mating. Chrysanthemum aphids overwinter in greenhouses where they feed throughout the winter. During the warmer months they leave the greenhouses in search of new plants. They also may be moved about as plants are shipped or sold. When the winged female stage infests new plants, it usually starts feeding and producing live nymphs. Each female can produce four to eight young aphids per day.

Within about a week the new nymphs mature into wingless females which begin to bear young of their own. One aphid on a plant in a short time may build the population up to hundreds of individuals. As the plant becomes crowded, more and more of the offspring develop into winged females which in turn migrate to other plants to begin new infestations.

Management Strategies

Infested plants in the greenhouse should be sprayed thoroughly when aphids are first noticed. On outdoor plantings, natural enemies may control minor infestations.

Green Peach Aphid

Scientific Name: *Myzus persicae* (Sulzer)

Class: Insecta

Order: Hemiptera

Family: Aphididae

Description

Adults: The small adult green peach aphid is light to dark green or pink, with red eyes. Three dark lines run down its back. Wings may or may not be present. The tobacco aphid is similar and can be either red or green.

Eggs: Found only in the northern United States, the egg is black and shiny for the green peach aphid. The tobacco aphid has not been found to have an egg stage.

Nymphs: The wingless nymph resembles the larger adult.

Biology

Host Plants: Green peach aphids have been collected from over 100 plants, including a wide variety of vegetable and ornamental crops. Spinach, potatoes, and peaches (the host on which eggs are laid) seem to be especially favoured hosts. Tobacco aphids will be pests primarily on tobacco and closely related plants such as flowering tobacco and *Nicotiana*. It is probably not a major pest, but will feed on pepper and eggplant as well as cole crops such as turnips, kale, and collards.

Damage: Aphids suck plant sap and contaminate the host with honeydew and cast skins. Some hospitals refuse to allow cut flowers in patients' rooms because of the mess by aphids. They are also the vectors of a number of plant viruses including tobacco, tomato, lettuce, dahlia, canna, and bee mosaics as well as tuber spindle, rugose mosaic, and leaf roll diseases of potato.

Life Cycle: In the northern United States, green peach aphids overwinter as eggs, but in the Southeast, no eggs are laid. Instead, female aphids give birth to young females during the growing season. The reproductive capacity of green peach aphids has been described as "fantastic." High reproductive rates and resistance to pesticides make the green peach aphid a formidable pest in the greenhouse. Up to 30 generations per year may take place in this pest's southernmost range.

The tobacco aphid probably overwinters on weed hosts or on cole crops that remain alive through the winter. Tobacco aphids are not known to have an egg stage, and they reproduce by giving birth to live young female aphids without mating. Their young are able to produce young as well without mating. If the plant becomes too crowded or if it becomes late in the season and the aphids need to find a winter host, the aphids give birth to young that will grow up to have wings and can move to other plants. Differences on reproductive rates exist between the red and green forms of the tobacco aphids. The red form is able to reproduce much faster during extremely hot weather than the green form.

On chrysanthemums, green peach aphids feed on all parts of the plant (melon aphids feed only on the buds and leaves, and chrysanthemum aphids feed only on the stems and leaves). Green peach aphids will not become established in the presence of the other two aphids unless pesticides are applied. In that case, green peach aphids outlive both melon aphids and chrysanthemum aphids.

Management Strategies

Because green peach aphids overwinter on weed hosts, infestations can occur in the greenhouse any time of year. Green peach aphids readily infest bedding plants and can be introduced into greenhouses whenever bedding plants are brought in from another grower. Although damage per aphid is often not serious, these aphids reproduce so rapidly that serious harm can be done in a short time.

Biological Control: Ladybugs, lacewings, syrphid flies, damsel bugs, wasps, and parasitic fungi tend to regulate green peach aphid populations outdoors.

Pesticides: These aphids' resistance to pesticides calls-for thorough applications whenever a new infestation is found. Tobacco aphids can be controlled in the same manner as green peach aphids.

Melon Or Cotton Aphid

Scientific Name: Aphis gossypii Glover

Class: Insecta

Order: Hemiptera

Family: Aphididae.

Description

Adults: This is a small aphid, smaller than most other aphids. The winged adults are about 1.25 millimetres long, soft bodied, and yellow to dark green with a black head and thorax. The wings are held rooflike over the abdomen at rest. Wingless adults tend to be 1.0 to 1.5 millimetres long, uniform in colour, and yellow to dark green. The antennae and cornicles are shorter than those of winged adults. Cornicles are small, tail-pipe-like structures on the end of the body. Pale individuals tend to be smaller and to have fewer antennal segments than dark individuals.

Nymphs: Nymphs resemble adult aphids except for size (about 0.5 to 1.0 millimetre long). Those destined to be winged adults have wing buds in the later instars.

Biology

Host Plants: Melons and other cucurbits, okra, hops, strawberries, beans, spinach, tomatoes, clover, asparagus, citrus, catalpa, violet, hydrangea, begonia, ground ivy, and weeds are some melon aphid hosts. They have been discovered feeding on plants in 25 plant families. The melon aphid is an important pest of cotton and is also called the cotton aphid.

Damage: The melon aphid feeds by piercing the plant surface with its threadlike mouthparts to suck out plant juices. This feeding causes distorted growth, decreased yield, reduced quality of yield, and prematurely ripened fruit. The fruit may be covered by the feeding aphids' honeydew and by cast skins.

The melon aphid transmits several important plant viruses including cucumber mosaic, onion yellow dwarf, citrus quick decline, lily symptomless diseases, and lily rosette.

Life Cycle: The melon aphid is an important pest of both agricultural and ornamental plants. Being practically omnipresent, it feeds upon many host plants. The melon aphid spends the winter on weed hosts and on cold-tolerant plants probably both as nymphs and adult females in the south. During warm periods of winter they start feeding until cold weather inactivates them again. In the spring the adult females move to new hosts and start feeding and rapidly reproducing.

In northern climates the aphid overwinters in the egg stage. Indoors and in greenhouses the aphids feed and reproduce throughout the winter. Melon aphids commonly start out on one plant and spread out from that point. On woody ornamentals such as gardenias, feeding is confined to new growth in the spring.

For the melon aphid there are two kinds of hosts, primary and secondary. In late fall, aphids feed upon primary plants mate and lay overwintering eggs. Melon aphids feeding on secondary plants always give birth to live young. In spring, winged forms usually infest new plants, both primary and secondary, and the females produce live nymphs.

Within about a week the new nymphs mature into wingless females that begin to bear young of their own. As the plant becomes crowded, more and more of the offspring develop into winged females which in turn migrate to other plants to begin new infestations. Wingless forms usually predominate in low aphid populations.

Management Strategies

Biological Control: Natural enemies may control minor infestations on outdoor ornamentals. Syrphid fly maggots and ladybird beetles and their larvae feed upon melon aphids. Braconid wasps parasitize the aphids, and ants feed on the honeydew excreted by feeding aphids.

Pesticides: Because the winged forms are 2 to 3.7 times more resistant to organophosphate pesticides than are wingless forms, infested plants in the greenhouse should be sprayed thoroughly when aphids are first noticed.

Mealybugs

One of the more common groups of scale insects attacking ornamental plants are called mealybugs. There are about 275 species of mealybugs known to occur in the continental United States. Mealybugs are prevalent pests in greenhouses and interior plantscapes such as shopping malls, conservatories, hotels, and office buildings.

Mealybugs cost growers and retailers millions of dollars per year in control costs and crop damage or loss. Damage is caused by mealybugs feeding on host tissues and injecting toxins or plant pathogens into host plants. In addition, mealybugs secrete a waste product, honeydew, which is a syrupy, sugary liquid that falls on the leaves, coating them with a shiny, sticky film. Honeydew serves as a medium for the growth of sooty mold fungus that reduces the plant's photosynthetic abilities and ruins the plant's appearance. Feeding by mealybugs can cause premature leaf drop, dieback, and may even kill plants if left unchecked.

Mealybugs are one of the more active groups of scale insects as most of them retain well-developed legs and remain mobile throughout their life. However, they generally move little once a suitable feeding site is found.

They are small insects (1 to 4 millimetres long) and the body is usually covered with a white cottony or mealy wax secretion. This makes them appear like small spots of cotton on the plant, particularly when the female is laying eggs and producing an ovisac to cover and protect the eggs. Mealybugs generally have an oval body outline. Many of them produce marginal filaments of wax that may be wedge-shaped or spine-like, but others lack marginal filaments entirely. As immatures, male and female mealybugs look alike, but as adults they are quite different. The adult male looks more like a small two-winged fly.

The life history of mealybugs varies depending on the species. Basically female mealybugs go through four developmental stages or instars and as adults may lay up to 600 eggs, usually in a cottony-like ovisac beneath her body. The eggs hatch in 6 to 14 days and the first instars or "crawlers", as they are commonly called, disperse to suitable feeding sites on new plant parts or hosts. They can survive only about a day without feeding, and once they insert their stylets to feed they generally remain anchored permanently.

The crawler stage is the most fragile and easily controlled stage in a mealybug's life history. Some mealybugs, like the longtailed mealybug, do not lay eggs, but bear their young as active crawlers. Male mealybugs go through five instars and feed only in the first two instars. Adult males have no functional mouthparts, live only a day or two, and exist solely to fertilize the females.

Outdoors most mealybugs go through one or two synchronized generations and overwinter as second instars. Indoors, there may be a continuous overlapping of generations and all stages can be found on the host at a given time. The citrus mealybug may have as many as eight generations indoors. The overlapping of generations makes control difficult. One of the first methods of control is to purchase plants that are not infested with mealybugs. Commercial flower growers sometimes discard plants infested with mealybugs rather than try to rescue them with insecticidal treatments.

Common mealybugs occurring indoors include: the citrus mealybug, the longtailed mealybug, the Mexican mealybug, and Pritchard's ground mealybug. Citrus mealybug is by far the most common and widespread mealybug pest. It attacks nearly every flowering species grown in the greenhouse.

Key To The Most Common Mealybugs Found On Flowers And Foliage Plants:

1. Longtailed Mealybug. Adult females with long, white filaments at the rear; females apparently giving birth to live young.

2. Female feeding on above ground portion of plant.
3. Citrus Mealybug. Body orangish or purplish and covered with white bloom; a darker line down the middle of the back; mealybug very damaging to the health of many plants.

Citrus Mealybug

Scientific Name: Planococcus citri

Class: Insecta

Order: Hemiptera

Family: Pseudococcidae.

Description

Adults: The female citrus mealybug is wingless and appears to have been rolled in flour (hence the name). It grows to 3 millimetres long and 1.5 millimetres wide. A fringe of small waxy filaments protrude from the periphery. The male is small, but with its wings and tail filaments, it appears to be 4.5 millimetres long.

Eggs: The oblong, yellow eggs are enmeshed in a dense, fluffy, white ovisac.

Crawlers: The tiny crawler is oval and yellow, with red eyes. The antennae are rather distinct.

Nymphs: Female nymphs resemble the larger adult females. Male nymphs are narrower and often occur in a loose cocoon.

Biology

Host Plants: Citrus mealybugs have been collected from at least 27 host plant families. Many ornamental plants grown in greenhouses are susceptible to attack including begonia, coleus, amaryllis, cyclamen, and dahlia. Citrus mealybug has been collected on canna, narcissus, and tulip outdoors.

Damage: Citrus mealybugs damage hosts by sucking out plant sap, by excreting honeydew in which sooty mold can grow, and by causing distorted growth and premature leaf drop with their toxic saliva. They further disfigure plants by secreting cottony wax. Infested plants usually die unless the pest is controlled.

Life Cycle: The citrus mealybug has been recognized as a pest of citrus and ornamental plants in Europe since 1813 (where it is called the greenhouse mealybug) and in the United States since 1879.

Because female citrus mealybugs have no wings, they must be transported to the proximity of the next host plant. They can, however, travel short distances by crawling. The immatures can be blown about. Males are small, winged insects.

After mating, each female lays up to hundreds of eggs in a dense, fluffy secretion called the egg sac or ovisac. Within a few days, new mealybugs (crawlers) hatch and begin to squirm out of the ovisac. Light infestations are easily overlooked because the mealybugs tend to wedge into crevices on the host plant. As their numbers increase, mealybugs of all sizes can be seen crawling around or feeding on all exposed plant surfaces.

Management Strategies

Control of citrus mealybugs is amazingly difficult. Some commercial flower growers merely discard infested plants rather than trying to rescue them from citrus mealybugs. Horticultural oils may damage amaryllis.

Biological Control: The lady beetle *Cryptolaemus montrouzieri* and parasitic wasps *Leptomastix dactylopii* and *Anagyrus pseudococci* attack citrus mealybugs.

Longtailed Mealybug

Scientific Name: Pseudococcus longispinus

Class: Insecta

Order: Hemiptera

Family: Pseudococcidae.

Description

Adult Females: Up to 3 millimetres long, female longtailed mealybugs have 17 pairs of waxy filaments around the periphery. On mature specimens, the caudal (tail) filaments are as long as or longer than the yellowish to grayish body (unless the tail filaments have broken off). Overall length may be 6 to 7 millimetres.

Eggs: There is no external egg stage of the longtailed mealybug.

Nymphs: The nymph is similar to the larger adult female except that the filaments around the edges are shorter.

Biology

Host Plants: Longtailed mealybugs have been found on at least 26 plant families. Dracaena appears to be the favoured host, but most flowering and ornamental foliage plants are susceptible.

Damage: Longtailed mealybugs feed by sucking out plant sap from leaves and stems. Honeydew and sooty mold further disfigure infested plants, which may eventually be killed. These pests also secrete a fluffy white wax which also detracts from the appearance of infested plants.

Life Cycle: Although longtailed mealybugs were first described in 1867, not much has been published on their biology. Females give birth to live young on a shallow pile of white waxy secretions. Because the females are wingless, they must be brought into proximity of a host plant before it can be infested.

Management Strategies

Biological Control: A small wasp, *Anagyrus nigricornis*, parasitizes longtailed mealybugs, and a small, predaceous, brown lacewing insect feeds on them. Ants sometimes protect longtailed mealybugs from parasites and predators and feed upon the honeydew excreted by these mealybugs.

Pesticides: Longtailed mealybugs are sometimes difficult to control, even though there is no egg stage protected by a dense ovisac. If many plants are infested, they should be dipped or sprayed thoroughly with a pesticide mixture or the entire greenhouse treated with an aerosol. It is best to retreat two or more times at weekly intervals. Retreatment will control mealybugs that were missed by earlier control efforts.

Mexican Mealybug

Class: Insecta

Scientific Name: Phenacoccus gossypii Townsend and Cockerell

Order: Hemiptera

Family: Pseudococcidae.

Description

Adults: The female Mexican mealybug adult is 3 to 4 millimetres long, oval, grayish and covered with a thin waxy secretion. There are three parallel rows of small waxy tufts down the back.

This insect is a short-tailed mealybug (the caudal filaments do not exceed 1/4 the body length). The lateral filaments are also short. Males are small gnat-like insects with only two wings. Adult Mexican mealybug males have four waxy, posterior filaments.

Eggs: The egg sac is white, dense, narrow, and longer than the female secreting it.

Nymphs: The nymphs are small and yellowish with white waxy secretions.

Biology

Host Plants: The Mexican mealybug is found commonly on numerous ornamental plants, a few of which are aralia, chrysanthemum, English ivy, geranium, Gynura, hollyhock, Ixia, lantana, and poinsettia. This insect is also a minor pest of lima beans in the warmer parts of the United States.

Damage: Wilting and stunting are common symptoms of Mexican mealybug attack. This insect can be as damaging as the citrus mealybug. The mealybugs and ovisacs also disfigure heavily infested plants.

Life Cycle: In the greenhouse, the Mexican mealybug may have seven complete generations in one year. The average time required from oviposition to the adult stage is 47 days. Each female deposits about 400 eggs. The eggs are enclosed in an elongate cottony mass called the ovisac that originates at the back of the female. The ovisac is about 6 millimetres long. The eggs hatch in 6 to 14 days. The life cycles of the male and female Mexican mealybugs differ. A female passes through three nymphal stages only. Male Mexican mealybugs pass through two nymphal stages and two resting stages (prepupal and pupal stages). Normally, these mealybugs are found above ground on the leaves, stems or flowers, but occasionally will be found feeding on the roots.

Management Strategies

Pesticides: For chemical control recommendations, consult the Cooperative Extension Service.

Root Mealybug

Scientific Names: Ground Mealybug: *Rhizoecus falcifer* Kunckel d'Hercularis; Pritchard's Mealybug: *Rhizoecus pritchardi* McKenzie

Class: Insecta

Order: Hemiptera

Family: Pseudococcidae.

Description

Adult Females: The ground mealybug is white and 2.4 to 3.9 millimetres long. It resembles a springtail, but moves much more slowly and cannot jump. The ground mealybug has slender waxy filaments that form a sort of netting over some individuals. The ground mealybug also secretes a small amount of wax, which can give the soil a somewhat bluish appearance when the mealybugs are abundant. Pritchard's mealybug is snow white and 1.6 to 2.1 millimetres long and oval. It has small to non-existent eyes.

Biology

Host Plants: The ground mealybug feeds on the roots of anemone, chrysanthemum, gladiolus, iris, and numerous other flowers, shrubs, and ornamental grasses. Pritchard's mealybug has become a serious pest of African violet, although it also infests *Achillea*, *Arctostaphylos*, *Geum*, and *Polygala*.

Damage: At times the ground mealybug becomes abundant enough to damage its host. Pritchard's mealybug causes devitalization, foliage deterioration, and even death of its host plant.

Life Cycle: Little has been published on the life history of root mealybugs. When infested African violets are irrigated, Pritchard's mealybugs crawl out of the drainage holes and spread throughout the greenhouse. Eggs are laid in a loose ovisac in clusters of at least six eggs. All stages can be found on the roots and potting mix of African violets where they resemble slow moving, snowy white collembola.

Management Strategies

Pesticides: Pesticides applied as dips, drenches, or granules are more effective for root mealybug control than are foliar sprays. For specific chemical control recommendations, consult the Cooperative Extension Service.

2

Physiology and Biochemistry of the Insect

Structures of Fat Bodies

The morphology of fat bodies varies considerably in various insect groups. In many species it is a loose aggregation of cells that may be unevenly distributed among all body segments, including the brain case. Fat body cells often surround the major organ systems in insects, including reproductive tissues, alimentary canals, thoracic muscles, and elements of central nervous systems. This close association of fat body cells and other tissues may facilitate exchanges of macromolecules.

Fat bodies may also protect organs from physical damage by providing packing. In other species fat bodies are compact cell masses invested in a membranous covering such that they easily can be removed from the insect intact. Structures of fat bodies may vary according to life stage, sex and reproductive status. Fat bodies can have different functions between sexes, and functions also can differ among life stages. In some species they undergo histological breakdown and reformation during papal stages, though in most species they do not appear to do so.

Fat bodies of all insects are in intimate contact with circulating hemolymph, which is consistent with movements of molecules between the two compartments. The vast majority of the cells in fat bodies are called trophocytes, and it is in these cells that most fat body functions are carried out.

Trophocytes are very similar to some hemocytes; the two cell types may be quite closely related. In the homopteran Aleyrodes, for example, fat body cells float individually in hemolymph and can not be clearly distinguished from hemocytes. Other fat body cell types include urate cells. These cells are associated with storage excretion, in which uric acid crystals are accumulated rather than excreted. Fat bodies of most insect species do not include urate cells. For the sake of complexity, uric acid is stored in trophocytes of some species. Fat bodies of many species also contain mycetocytes, which are involved in housing symbiotic microorganisms. Many of these symbiants play crucial roles in insect nutrition by providing certain vitamins and other essential nutrients to insects that are not able to acquire these nutrients in the diet.

Protein Biosynthesis in Fat Body

Many of the proteins that are crucial in the lives of insects are biosynthesized in fat bodies. Protein synthesis is often under control of juvenile hormone (JH). One of the most well-studied examples of endocrine regulated protein synthesis in fat bodies is the formation of vitellogenins by adult females. Vitellogenins are the yolk proteins that are accumulated during egg maturation. Fat bodies biosynthesize many other proteins as well, including virtually all of the major proteins found in the hemolymph of larval insects.

There are also many examples of specialized proteins in insects. Larvae of the midge *Chironomus* biosynthesize hemoglobins, heme-containing proteins important in storage of oxygen. Hemoglobins can make up more than 80% of the hemolymph proteins in these larvae. The lipoproteins that function in hemolymph transport of diacylglycerol are synthesized in fat bodies. There are many other examples, all bearing on the point that protein synthesis in fat bodies is a central issue in insect biology.

We review the biochemistry of protein synthesis by way of appreciating the importance of this topic. First we recall the expression originally put forth by Francis Crick, and later articulated by Charlie Brown as the central dogma of molecular biology: DNA MAKES RNA MAKES PROTEIN. As an aside, a group of my biochemical colleagues at the University of Nevada, Reno and myself once entered a biochemical T-shirt logo contest. We added the expression makes lipids to the expression of the central dogma of molecular biology. We did not win. This figure shows that genes are transcribed from DNA to RNA within a cell nucleus.

The transcription process produces a special form of RNA called messenger RNA, or mRNA. Messengers go places, and in this case mRNA makes its way from the nucleus to the cytosol where it joins with ribosomes, modelled in the figure. We want to see that the mRNA is joined with two ribosomal subunits, a small and a large subunit.

The flow of information is from the nucleus to the cytosol. This is followed by a series of steps that include translation of the genetic information borne on the mRNA strand into a polypeptide chain, or string of amino acids. The key here is that each of the amino acids can be attached to a specific RNA molecule known as transfer RNA, or tRNA. One portion of tRNA molecules feature an anticodon that matches up to codons on the mRNA strand. Ribosomes have sites for two tRNA molecules, one called a peptidyl site and the other called an aminoacyl site. While two tRNA molecules are juxtaposed within the ribosome a peptide bond is formed between the amino acids attached to each tRNA.

At this point the tRNA molecule in the peptidyl site is released and the ribosome scoots one step along the mRNA strand, thereby shifting the tRNA molecule in the aminoacyl site to the peptidyl site and opening a space for a new tRNA molecule. The cycle goes on with addition of the new tRNA molecule, formation of another peptide bond, release of a free tRNA molecule and another ribosomal shift. A copy of the universal genetic code is appended to the end of this column. Please note, for example, that the AGG in the mRNA strand shown above specifies the amino acid arginine.

Protein synthesis begins with transcription of a DNA strand within a cell nucleus. JH expresses its effects on protein synthesis within fat body cells by interacting with DNA. JH receptor proteins are found within the cytosol of cells. JH forms a complex with its specific receptor, then the hormone-receptor complex moves into the nucleus. It is the JH-receptor complex that interacts with DNA.

Carbohydrate Metabolism in Fat Body

The natural foods of many insects include substantial amounts of carbohydrates, or sugars, either in the form of simple sugars or as polymers. Following absorption across midgut epithelium, sugars are removed from circulating hemolymph into fat bodies. Sugars are assembled into large macromolecules of glycogen and stored within fat bodies (try to imagine a discussion on the physiology of the sugar body). The polymerization of simple sugars into glycogen is called glycogen synthesis, and is carried out enzymatically by glycogen synthetase. It has been suggested that glycogen synthetase may be under control of a sort of hypoglycemic hormone. There are a number of lines of evidence for factors that reduce hemolymph sugar levels, but the specific points of control remain unknown.

But there is also evidence for endocrine regulation of carbohydrate mobilization. Various names have been applied to hormones that increase hemolymph sugar levels, including hypertrehalosemic factor and glucogon-like hormone. Most recently people have been using the term trehalogon by analogy to the mammalian hormone GLUCOGON. I think such words give insect endocrinologists a bad name, but there you have it. Trehalogon is thought to stimulate fat bodies to release trehalose into circulating hemolymph.

Releasing trehalose into hemolymph is a multi-step process that begins with hydrolysis of individual glucose residues from glycogen. This next figure shows this is under control of the enzyme phosphorylase, so named because phosphorus is involved in the hydrolysis. This enzyme yields molecules of glucose-1-phosphate, which are taken into two pathways, one is a mutase step to glucose-6-phosphate and the other is a phosphorylase step to uridine diphosphoglucose. These two glucose derivatives are combined into trehalose-6-phosphate by trehalose-6-phosphate synthetase. A phosphatase step yields the free trehalose molecule. Hence, trehalose synthesis requires hydrolysis, derivatization and assembly of two glucose residues.

Trehalose synthesis is regulated by controlling the activity of phosphorylase, the enzyme at the beginning of the pathway. There appears to be two forms of phosphorylase, phosphorylase a, the active form and phosphorylase b, the inactive form. Both of these forms exist in fat bodies, and the specific ratio of one form to the other depends upon the physiological requirements of the whole organism. Physiological requirements are expressed by circulating titres of trehalogon, the hormone active in mobilizing fat body trehalose. A final point is a reminder that the figure as shown here is taken from a recent compendium on the endocrinology of insects. It contains a slight error, shown in the phosphorylase step. This step requires the input of inorganic phosphorus, just the opposite of what is shown.

Metabolism of Lipids in Fat Body

We have seen that lipid reserves are stored in fat bodies as triacylglycerol. Triacylglycerol is synthesized in fat body from excess sugar and from the carbon skeletons of certain amino acids by way of the ubiquitous enzyme system fatty acid synthetase. Newly formed fatty acids are linked to glycerol backbones by acyltransferases.

Lipid is released from fat body in the form of 1,2-diacylglycerol (DAG), rather than as 1,3-diacylglycerol or a mixture of 1,2-DAG and 1,3-DAG. 1,2-DAG may be formed directly by a lipase specific to the 3-position fatty acid, or it may be formed by breakdown to 2-monoacylglycerol followed by stereospecific reacylation to 1,2-DAG. Release of 1,2-DAG is stimulated by adipokinetic hormone, but the particular enzyme target is not yet known.

Uric Acid Synthesis and Other Detoxification Reactions

The details of uric acid synthesis will be covered in the section on nitrogen metabolism and excretion. The point to emphasize here is that the fat body is the central organ of nitrogen metabolism. The storage of uric acid in urate cells and in trophocytes by some insects has been covered.

One of the great problems in crop protection by application of chemical pesticides is the natural resistance to these poisons that insects rapidly develop. Two broad mechanisms of resistance are enzymic breakdown steps, one by esterase and another by mixed-function oxidases. Under these systems, the pesticide molecules are simply metabolized into harmless forms before they can carry out their damaging effects. Most of these detoxification enzymes are formed in fat bodies and function in circulating hemolymph as well as in fat body cells.

Biosynthesis of Amino Acids in Fat Body

The non-essential amino acids can be formed using carbons from various intermediates in carbohydrate metabolism coupled with transaminations from preexisting amino acids. The fat body is the major site of amino acid synthesis.

At Least One Hormone is Synthesized in Fat Body

Most homeostatic hormones, such as diuretic hormone, adipokinetic hormone and trehalogon are released from elements of corpus cardiacum or from elements of neurosecretory cells. There is an example of a hormone that is produced and released from the fat body. About 40 hours after taking a blood meal, adult female mosquitoes produce what is tentatively called host-seeking behaviour inhibitory factor. This factor is a small peptide, and is produced in fat body and released into circulating hemolymph. The exact site of action remains to be found, but one physiological effect is to switch off specific lactic acid receptors located in antennae. Fat bodies need to be physiologically prepared to synthesize the factor. Another hormone appears to be required because females that have been ovariectomized within three hours after a blood meal do not elaborate the factor.

Energy Metabolism

Muscle contraction, ion transport, anabolic reaction pathways, and most other forms of cellular work are all powered by hydrolysis of adenosine triphosphate (ATP). The phosphodiester linkages in ATP are high energy bonds, and the potential chemical energy stored in ATP is released during the enzymatic hydrolysis of these phosphate bonds. The released energy is captured and utilized to power cell work. ATP is nearly ubiquitous, and is generally regarded as a sort of universal currency in cellular energetics. ATP is an unstable molecule, and large amounts can not be accumulated in anticipation of cellular activity. Hence, ATP is often generated during activity, especially energy-intensive activities such as insect flight.

ATP is part of the adenylate system. Adenosine monophosphote (AMP) and adenosine diphosphate (ADP) complete the system. These three molecules, AMP, ADP and ATP, store and transduce most of the energy used in cellular metabolism. The relative amounts of each make up what we call the "energy charge" of any cell.

If all adenosine in the adenylate system were present as ATP, the energy charge would be 1.0; if all adenosine were present as AMP the energy charge would be 0.0. Cells generally maintain a homeostasis of energy charge. When the energy charge declines below a set point appropriate to any given cell, there follows a shift toward ATP-generating pathways. On the other hand, when the energy charge is high, cells favour ATP-consuming pathways. Homeostasis of energy charge results from balancing ATP-generating and ATP-consuming pathways. Just to give you a number, the energy charge of most cells is about 0.85, perhaps something higher than we might expect. Let us wonder in a little bit more detail how cells balance ATP-generating and ATP-consuming pathways. We gain a bit more understanding by regarding the biological significance of the adenylate system. The adenylate system influences cellular metabolism in two ways. First, as just mentioned, the relative amounts of each adenosine species determines the energy charge of cells. Second, key enzymes in some metabolic pathways are regulated by the instantaneous concentrations of ATP and AMP. Hence, the adenylate system is both a regulatory and energy-yielding entity.

Respiration, or gas exchange, provides oxygen to mitochondria to support the rapid generation of ATP during energy intensive activities, such as flight. Mitochondria, and to some extent, extra-mitochondrial cellular enzymes generate ATP by the step-wise oxidation of chemical fuels. Because the oxidation takes chemical fuels through several intermediary steps, this process is often called intermediary metabolism. This subject could easily occupy us for the rest of the semester, so I here emphasize that this is one of many lectures in physiology-not a course in biochemistry.

We proceed with a series of questions. First, what are the flight fuels in insects? There are three major flight fuels. Honeybees and bumble bees, for examples, oxidize carbohydrate almost exclusively during flight. The most common carbohydrate for bees is the simple sugar glucose.

Some butterflies and moths rely for the most part on lipids, commonly known as fats, to support flight. Other insects, such as the tsetse and the Colorado potato beetle mostly re;u on one of the 20 amino acids-proline-during flight. Most insects are probably capable of using more than one energy substrate during various of their activities. For example, the desert locust usually begins a long flight powered by glucose metabolism; after about 15 to 20 minutes there is a shift to lipid metabolism for the duration of a long migratory flight.

Knowing the basic fuels used by most insects, we can next ask what are the usual storage forms of metabolic fuels? Carbohydrates are stored as glycogen. Glycogen is nothing more than a string of monomeric glucose units hooked up in a chain. Branches occur about every 8 to 12 glucose units in a glycogen macromolecule. The branching provides additional sites for rapid metabolism of glycogen: it can be elongated by adding simples sugars, and rapidly shortened by hydrolyzing them. The branching also increases solubility of the macromolecule.

In vertebrate liver glycogen collects into dark-staining granules. Such granules have also been seen in preparations of insect fat bodies. Here we note that the fat body is not, as the name suggests, simply a fat storage depot. Rather, it is a complex metabolic centre more comparable to the vertebrate liver. Most glycogen is stored in the fat body and to a lesser extent is found in flight muscle cells.

The next major fuel-lipid-is stored as triacylglycerol. This is made of a glycerol backbone, which is a 3-carbon unit. The backbone may have 1, 2 or 3 fatty acids bonded in ester linkage. The nomenclature, then, is monoacyl-, diacyl-, and triacylglycerol, depending upon the number of fatty acids bonded in ester linkage to the backbone. Most lipid is stored in the fat body, although small amounts are found in other cells.

Some insects oxidize the amino acid proline, sometimes expressed in the correct nomenclature as an imino acid. Insect hemolymph is quite high in amino acid content-much higher than in mammalian blood. The hemolymph is the major storage site of proline. Proline is unique among the energy substrates because its storage, transport and metabolic forms are all the same.

Given that we have some idea of the major metabolic fuels and their storage sites and forms, we can go on with the observation that, by and large, the fat body is the major site of energy storage. The fat body is relatively distant from the site of utilization inside flight muscles and other insect tissues. This implies that the fuels must be transported. During flight, for example, fuels are transported mainly from the fat body, found mostly in the abdomen, to the thoracic muscles. This is one of the major physiological roles of hemolymph: the transport of nut-rients. The next general question, then, is what are the major transport forms of chemical energy?

Again, we start with carbohydrates, generally stored as glycogen in the fat body. Glycogen can be shortened, one glucose unit at a time. The released glucose units can then be coupled into a molecule of TREHALOSE. Trehalose is the principle blood sugar in insects, and it plays a comparable role to blood glucose in mammals. Trehalose is a disaccharide of two glucose units. Once formed, it is released from the fat body into the hemolymph. In the hemolymph of blowflies, for an example, trehalose can be as concentrated to as much as 20 mg/ml. For comparison, human blood sugar is usually about 100 mg/decaliter or 1 mg/ml. Due to the high hemolymph concentration, trehalose diffuses from the hemolymph into the muscle cells and other tissues where it is hydrolysed into 2 glucose units. The high concentration of trehalose in the hemolymph suggests that upon closer inspection there are two storage forms and two storage sites for carbohydrates. Glycogen is stored in the fat body and trehalose represents another, hemolymph, storage form.

The physiology of lipids, which we saw to be mainly stored in the fat body as triacylglycerol, is somewhat different. Unlike sugars, which are soluble in rather high concentrations in hemo-lymph, lipids are functionally defined as being insoluble in aqueous media. Because of this physicochemical property, the predominant lipid transport form is as diacylglycerol associated with a specific hemolymph lipoprotein called *Lipophoren.* The most recent thinking, in the absence of real knowledge, is that fat-ty acids are hydrolysed from the glycerol backbone at the muscle cell membrane, then taken into the muscle while the lipophoren and glycerol backbone are recirculated back to the fat body.

We have come to the point of seeing what the major insect metabolic fuels are, their storage and transport forms and the main depots of storage. For the most part, energy metabolism in insects is not much different from other animals. A couple of major differences are transport of diacylglycerol and use of trehalose as a blood sugar.

	Carbohydrate	***Lipid***	***Amino Acid***
STORAGE SITE	Fat body/blood	Fat body	hemolymph
STORAGE FORM	Glycogen/trehalose	Fat body	Imino acid
TRANSPORT FORM	Trehalose	diacylglycerol	Imino acid
METABOLIC FORM	Glucose	Fatty acid	Imino acid

The next questions are what are the major metabolic pathways for each of these forms of chemical energy? And, where do the enzymes in these pathways occur? These questions are best taken together.

Glucose is broken down to pyruvate, or pyruvic acid, by way of the *Glycolytic* or *Embden-meyerhof* pathway. Depending on what biology or biochemistry text you happen to look at, there is a lot of noise in the representation of this pathway, so we will focus on the key features. These are that glucose enters the pathway, and pyrvate is the end product. There is a net gain of two ATPs per glucose molecule. In short, glucose is a 6-carbon sugar that gets broken down into two 3-carbon compounds. The enzymes of this pathway occur in the cytoplasm of cells, outside the mitochondria. In mere point of interest, speaking in terms of evolution, this is probably the oldest energy-yielding pathway, and there are numerous life forms that rely solely on this system.

Much more energy, a total of 34 ATPs, depending on how you add them up, is obtained from the complete oxidation of glucose to carbon dioxide and water. This more complete breakdown is accomplished by enzymes of the *Citric Acid* or Kreb's Cycle. The key features are (one) that acetyl Coenzyme A, formed by the decarboxylation of pyruvic acid, enters the cycle by joining the 2-carbon acetyl unit to a 4-carbon compound to create the 6-carbon citric acid and (two) there is no real end point because of the cyclic nature of the pathway. Rather, carbon dioxide is created at two points in the cycle by decarboxylations-recall that in organismal respiration carbon dioxide is removed by gas exchange-and electron carriers become loaded with electrons at 4 points in the cycle. Finally, the equivalent of 1 ATP is generated with each round of the cycle. The enzymes for the decarboxylation of pyruvate to acetyl CoA and the enzymes of the citric acid cycle are all located in the matrix of the mitochondria.

The final step in the complete oxidation of glucose is that the loaded electron carriers dump their electrons onto the *Electron Transport Chain*. In this system, electrons flow from the carriers, NADH and FADH2, through a series of electron carriers, finally combining with oxygen to form water. At three points along this chain a single ATP can be created for each electron pair that flows down the chain. Now, turning back to the Kreb's cycle, we can see that each turn of the cycle yields 12 ATPs-three each for the NADH (9), two for the FADH (11) and one for the GTP (12). Thus, Kreb's cycle and the ETC operate in concert to complete the oxidation of sugar to carbon dioxide and water, with generation of considerable amounts of ATP. Whereas the enzymes of the citric acid cycle are located in the matrix of the mitochondria, the ETC is buried within the inner mitochondrial membrane.

Let us consider the oxidation of fatty acids. Once transported across the cell membrane into the cell, the fatty acids are hooked up to CoA, making fatty acyl Coenzyme A, which can be transferred to fatty acyl carnitine. This is simply a transfer step to get the fatty acid into a form that can be transported into the matrix of the mitochondria where it is again hooked up to CoA. The carnitine, then diffuses back out of the mitochondrial space where it can reserve as a fatty acid carrier. Within the mitochondrial matrix, the fatty acyl CoA enters what is known as the ß-oxidation cycle. This pathway runs in circles, like most of us, and essentially repeatedly breaks 2-carbon units off the long-chain fatty acid which become acetyl Co-enzyme A. The acetyl CoA subsequently enters the Kreb's cycle and ETC where it undergoes complete oxidization to carbon dioxide and water. Recalling that one turn of Kreb's cycle yields 12 ATPs in concert with the ETC, we see that the 8 2-carbon units from a 16-carbon fatty acid can provide a considerable amount of energy for cell work.

The final substrate in our story will be the amino acid, proline. In the tsetse proline is oxidized in the sarcosomes, or mitochondria, according to this expression:

$$\text{PROLINE} + 21/2\ O_2 + 14\ \text{ADP} \rightarrow \text{ALANINE} + H_2O + 2CO_2 + 14\ \text{ATP}$$

In essence, the nitrogen is taken off of the skeleton and the carbon portion is oxidized to carbon dioxide and water by entering the Kreb's cycle at a point beyond citric acid.

Let us have a look at the biological significance of sugars, lipids and proline as metabolic substrates. With regard to pro-ine, we really have two questions. Why use an amino acid at all when it can be converted to carbohydrate which combines high solubility, good energy yield and convenient storage in hemolymph as trehalose and in fat body as glycogen? We understand proline metabolism best in the tsetse where it is part of the routine diet due to its blood-feeding habit. Bearing in mind that each step of any metabolic conversion results in some lose of energy, the idea is that it is more efficient to oxidize the dietary proline directly rather than convert it to something else. This is because the amino nitrogen atom must itself be metabolized through interconversion of other amino acids. It turns out that these other compounds are themselves better candidates for conversion to sugar. Then our second question would be why oxidize proline as opposed to some other amino acid? There are two main points.

One is the net yield of ATP per gram amino acid and the other is the solubility of the amino acids in aqueous hemo-lymph. On the basis of energy yield, probably only isoleucine and leucine are serious rivals to proline, with a possible argument for valine. But of these four amino acids, the solubility of all but proline is too low for maintenance of high concentrations with hemolymph. This argues that proline is probably the most suitable amino acid for energy substrate. And, again, may be quite appropriate when it is high in the diet.

With respect to flight duration, animals that can oxidize lipid as a flight fuel are capable of making long-range flights, whereas those that utilize carbohydrate are capable of only relatively short periods of continuous flight. This follows from the higher energy content of a given weight of lipid compared to sugar. Also, glycogen, the main carbohydrate store, is always stored with an appreciable amount of water associated with it. We can calculate that 1 mg of lipid can yield as much metabolic energy as about 8 mg of stored glycogen: thus, for long flights the weight economy of a fuel becomes important. Also, lipid produces more metabolic water than does carbohydrate and this may be important in the water relations of some insects. This reasoning is not limited to insects. Many migratory birds similarly accumulate lipid reserves to fuel long distance flights.

On the other hand, carbohydrate offers advantages to creatures like bumblebees. A key advantage is that the energy from glucose can be realized much faster than the energy from a fatty acid. It is generally true that the most rapidly metabolizing insects oxidize sugar.

Flight Muscle Physiology

The ability to move is one of the basic characteristics of living things. To make this point, let us consider a number of different sorts of movement. Most of us are intimately familiar with more or less large scale movements, such as walking or running or watching birds and insects fly. We are also aware of a good number of more subtle movements, such as our heart-beats and movements of our viscera, actions that occur within organisms. Movements also occur at the cellular level. Unicellular organisms are able to move themselves around, by amoeboid movement or the actions of flagella and cilia. Flagella and cilia are also found in certain cells, sperm for example, within metazoan animals. Movement takes place within cells, as well. Chromosomes are moved within cells by contractile elements of a spindle apparatus. Nutriments and other materials are transported within cells, down the length of a nerve axon, for example.

All movements are driven by variations on two general systems of proteins. One is the tubulin-dynein microtubular system, mostly seen within cells, and the other is the actin-myosin filament system of muscles. From the point of view of a comparative physiologist, muscle, and its actin-myosin contractile elements is a special tissue because no other tissue so clearly illustrates the idea of a common mechanism with so very many variations adapted to specific functions.

Turning our attention to insects, one of the most obvious features of insects is that most of them are able to fly. The common names of insects reflect flight again and again, for examples, dragonflies, damselflies, houseflies, mayflies, stoneflies, fruitflies, hoverflies, butterflies. I am sure you can think of many more examples. All bespeak the point that most insects are fliers. We can fairly say that most of the world's animals fly.

Flight bears on nearly every aspect of insect biology. Foraging for food, searching for mates, escape from predators or fly swatters, avoiding harsh or unfavourable conditions, and locating egg-laying sites are just a few points of insect biology intimately connected to flight behaviour. It follows that understanding various behavioural, aerodynamic, physiological, biochemical and ecological aspects of insect flight is valuable in understanding the nature of insects.

We first consider structures of insect flight muscles. Flight muscles (take a look) are arranged into three broad categories based on structural grounds. In tubular muscles, fibrils are arranged radially about a column of sarcoplasm that contains several nuclei. Mitochondria are interspersed in the radial area of the muscle. Dragonflies feature this type of flight muscle structure. Tubular muscles occur in the legs of most insect species. Close-packed muscles are found in Lepidoptera and Orthoptera; people who get their security from scientific idiom call these microfibrillar muscles. In close-packed muscles large mitochondria are dispersed between the cylindrical fibrils, which are about 1.5 to 2.0 μm in diameter. Fibrillar muscles are characteristic of Hymenoptera, Diptera, Coleoptera and true bugs. These muscles have large fibriles, up to 5 μm in diameter, with huge mitochondria and nuclei scattered about.

Even though different phylogenic groups have developed various structural types of flight muscles, there are major similarities among them. All the muscle types have fibres as their physiological units because the fibres are bounded by a continuous sarcolemma or muscle cell membrane with the basement membrane attached. There are descriptions of fibres composed of several cells, however, a fibre is more correctly a single multinucleated cell, or syncitium. Also, the contractile proteins within the fibrils appear to be in similar arrangement.

Now, before we go forward, let us take a minute to clarify some language. People began looking closely at muscles during the 19th century, before articulation of a general cell theory. A consequence of this is a history of set of parallel nomenclatures, one for muscles and another for most other cell types. The terms for muscle usually include the prefix sarc-or my-, from Greek for flesh and muscle. So what we usually call the cell membrane or plasmalemma is called the sarcolemma in muscle. Similarly, mitochondria are called sarcosomes, the endoplasmic reticulum is called sarcoplasmic reticulum and cytosol is referred to as sarcoplasm.

Still on nomenclature, muscles are comprised of fibres, which are made up of fibrils within which are the contractile filaments. The ultrastructural arrangement of muscle is as follows: filaments are made up of the contractile proteins within muscle fibrils. Muscle fibres contain groups of fibrils; muscles are made of groups of fibres.

The most widely accepted mechanism of movement is known as the sliding filament theory of muscle contraction, first proposed in 1954. Each contractile unit is called a sarcomere, bounded on either end by Z-lines or discs. The area between Z-discs is sometimes called a Z-band. Other landmarks are the A-bands, the H-zone and the I-band (take a look). These bands are visible in microscopic preparations and are associated with the presence or absence of the major contractile proteins, actin and myosin. These terms date back to very early microscopic observation of muscles. The A of A-band stands for ANISOTROPIC, and the I of I-band stands for ISOTROPIC.

The isotropic bands have uniform indices of refraction in microscope preparations. Consider a cross-sectional representation of muscle fibrils where thin actin filaments are arrayed around thicker myosin filaments. Try to get into the habit of saying thick myosin and thin actin. It may be easier to keep the names straight that way. In flight muscles, the actin filaments are very often found in a hexogonal array around the myosin filaments while in slow twitch visceral muscles the actin is in a circular arrangement. A last point here is that all insect flight muscles are striated. Variations in the register of the Z-discs within a muscle make this feature more or less apparent in muscles from insects of different orders.

In addition to actin and myosin filaments, other proteins are present in relatively small quantities. These include troponin and tropomyosin, collectively called native tropomyosin. The native tropomyosin is involved in regulation of the muscle contraction.

We know, now, that the physiological unit of contraction, the muscle fibre, is invested by the sarcolemma, or functional cell membrane. At frequent intervals this membrane is invaginated to form a system of transverse T-tubules. The invaginated transverse T-tubules lie in close association with vesicles of sarcoplasmic reticulum. Additional invaginations of the sarcolemma allow tracheoles to be very close to the sarcosomes which consume amazing amounts of oxygen during in-flight energy production. Note also that still more invaginations allow motor axons to have multiple synapses with the muscle fibre. This is an important point because unlike vertebrate muscles, the muscle membranes of insects does not propagate action potentials very. There is a requirement, then, for synaptic junctions about every 100 μm or so.

Summarizing muscle structure, the three structural types of muscle, tubular, close-packed and fibrillar, have in common the basic cellular organization of contractile proteins that present a more or less visible striation. Various invaginations for neuromuscular junctions, transverse T-tubule systems and tracheole position appear in all muscle membranes. We can now move on to physiology of muscle contraction. A motor axon impulse is transmitted across a neuromuscular synapse to the muscle cell membrane resulting in a depolarization which spreads to the interior of the muscle fibre by the transverse T-tubule system. This is very well developed in insect flight muscles. Depolarization of the transverse T-tubule system causes release of calcium from vesicles of the sarcoplasmic reticulum. The calcium activates myosin ATPase, resulting in the two type of filaments sliding past each other. What we want to see is that neither the thin actin filaments nor the thick myosin filaments shorten during contraction. Rather, the Z-bands become increasingly shorter as the two types of filaments overlap each other more and more.

There is a great deal of experimental evidence on muscle contraction. If calcium is removed from an *in vitro* preparation of locust leg muscle, for example by addition of EGTA which selectively removes divalent cations, the muscle can not be stimulated to contract. Adding calcium chloride back into the buffer restores the ability to contract. Furthermore, calcium is resequestered by vesicles of the sarcoplasmic reticulum in the presence of ATP. The requirement for ATP shows that uptake of calcium requires energy: that is, it takes energy to relax!

The role of calcium in the interaction between thin actin and thick myosin filaments that result in muscle contraction is modulated through native tropomyosin. Troponin inhibits the actin-myosin interactions. This inhibition is calcium dependent due to the presence of tropomyosin. Thus, the contractile process involves an impulse from a motor axon which depolarizes the post-synaptic muscle cell membrane; the depolarization sweeps inward by way of the transverse T-tubules, resulting in release of calcium from vesicles of sarcoplasmic reticulum. Interaction of calcium and tropomyosin, which has specific calcium receptors, temporarily breaks the inhibition of actin-myosin interaction by troponin so that the two main contractile proteins slide past each other with the concomitant expenditure of ATP. The contraction cycle is concluded with the repolarization of the muscle cell membrane, the resequesterization of calcium by sarcoplasmic reticulum and relaxation of the muscle. Please note that relaxation is not the same thing as stretching the muscle to its uncontracted length. Most skeletal muscle systems in vertebrates and invertebrates occur in antagonistic pairs: following relaxation, a flight muscle is restretched by its pair-mate.

Let's go back to the idea that the two main contractile proteins slide past each other with the expenditure of ATP. We want to consider how the hydrolysis of ATP is translated into directional movement.

We begin by appreciating a few points about myosin. Thick myosin filaments have a certain springiness or elasticity. Within a single filament, this allows some myosin heads can be attached to thin actin filaments while others are not attached. Each myosin filament has something in the range of 500 myosin heads, and each myosin head is able to bind and hydrolyze ATP. The process of actin and myosin filaments sliding past each other results from the combined actions of all the myosin heads.

Myosin heads move actin filaments in a cyclic process. A free myosin head can bind a molecule of ATP, the immediately hydrolyze the ATP into ADP and inorganic phosphate. Both products remain bound to the myosin head, and the energy released by hydrolyzing the ATP forces the myosin head into a highly strained conformation. The energy from the ATP is stored in the strained myosin head. The strained head can then weakly bind to a subunit of an actin filament. The weak interaction between the actin and strained myosin results in the release of the inorganic phosphate. The strained myosin, with ADP still bound, then binds very tightly to its actin subunit. Once tightly bound, the myosin head undergoes a conformational change that pulls the actin filament. This is the "power stroke", at the end of which ADP is released from the myosin head. The myosin head can then bind another molecule of ATP and release the actin. The free myosin can then hydrolyze the ATP, and the cycle continues. A myosin head may go through about 5 contraction cycles per second.

The key point here is that the hydrolysis of ATP is not directly coupled to the power stroke. Rather, the energy from ATP hydrolysis is stored briefly in a strained protein conformation and subsequently released. This is a theme running through many aspects of cell biology. Ion pumps and multi-enzyme macromolecule synthesizing systems similarly capture energy in strained protein conformations.

In addition to placing muscles into three groups based on structure, flight muscles are grouped into two sorts based upon a functional relationship between nerve firings and muscle contraction. In synchronous flight muscle systems, there is a one-to-one correspondence between each nerve impulse and subsequent muscle contraction. Synchronous flight systems are also called neurogenic systems and include Odonata, Orthoptera and Lepidoptera. In asynchronous flight systems, a single nerve impulse gives rise to several muscle contractions. Asynchronous systems are also known as myogenic systems because the stretch of one set of muscles by their antagonistic pair mates stimulates contraction: the contraction is caused by the muscle, myo, rather than nerve, neuro. Diptera and Hymenoptera are examples of myogenic fliers.

The synchronous flight muscle systems are associated with tubular and close-packed flight muscles, and are characterized by relatively slow wing-beat frequencies. Butterflies produce about 4 to 20 beats/second and locusts about 15 to 20. On the other side, the myogenic systems are associated with fibrillar muscles and quite high wing-beat frequencies. Bees and some flies, for examples, operate at about 190 beats/second, with extremes of up to about 1000 beats/second in some very small flies. The highest frequency recorded is 2218 Hz in the midge Forcips , tested at high temperature and with the wings clipped.

At first inspection, the contractile process as we understand it may not seem consistent with such high frequencies. Let us try to reconcile this. In both neurogenic and myogenic systems the muscles are attached such that a very short contraction permits the full excursion of wing movement. This is more extreme in myogenic systems. For example, in Sarcophaga flight muscles contract by only about 1% of their length during flight.

Myogenic muscles can be made to rhymetically contract if the muscle is stretched and set into oscillation by a mechanical device. Stretching the fibrile produces a slightly delayed contraction, and releasing the tension produces a slightly delayed relaxation. The muscle works by adding energy to maintain an already oscillating system. In the whole animal the thoracic box and wings together form a resonating system, and the timing and duration of muscle contraction is mechanically induced. The operation is started by a direct nervous command; after that first impulse, further nervous input regulates the availability of calcium which in turn regulates the power output of the mechanical system by way of the regulatory proteins we mentioned, that is, native tropomyosin. The fact that the muscle is stimulated to contract by stretching rather than by a nervous impulse gives rise to the term myogenic.

In synchronous fliers, clipping the wings has little effect on frequency because the basic frequency is programmed into the thoracic ganglia. In flies and other myogenic fliers, however, clipping the wings nearly doubles with wing beat frequency without a change in membrane potential spike frequency. These muscles can not be driven by electrical stimulation to go as fast as their normal frequency.

Again, this points to the mechanical nature of the asynchronous flight muscle system. In the whole animal the muscles are not attached to the wings, but to particular elements of the thorax such that the wings are part of a lever with two stable positions-up and down. When muscle contractions move one arm of the lever beyond the mid-point, the skeletal elements spring in, which shortens the muscle and releases tension. But at the same time this motion stretches the pair-mate, causing it to contract. The contraction moves the lever in the other direction, and tension is redeveloped as the lever reaches the mid-point. This oscillating cycle goes on as long as the muscle is maintained in an active state by a relatively small number of nerve impulses. In the presence of calcium, which is released from the sarcoplasmic reticulum due to nerve impulses, myosin ATPase splits ATP during oscillatory work. Stretch itself increases the activity of calcium-activated myosin ATPase-here stretch acts like an increase in calcium. This is also consistent with differences between myogenic and neurogenic systems. In neurogenic systems, each wing beat may be associated with a contraction-relaxation cycle, with substantial movement of calcium out of and back into the sarcoplasmic reticulum. Asynchronous muscles are less sensitive to calcium flux, which is also correlated with reduced development of sarcoplasmic reticulum.

Water Balance in Terrestrial Insects

We have spent a little bit of time considering aspects of movement. We've gotten an idea of the contractile process in muscle, a process driven by hydrolysis of ATP. We saw that ATP is generated as an end product of energy metabolism and considered gas exchange as a requisite for energy metabolism. Flight is very important to many aspects of insect biology, and I purposefully emphasize flight. Let us also emphasize the broader biological issue in flight. Flight can be taken under the mantle of motility, or movement. In this view, flight is one of many kinds of movement, all fundamental processes in biology.

Let us now turn attention to still another are of physiology confronting nearly every life form: osmotic balance. Osmotic regulation is the business of maintaining optimal salt and water concentrations in an organism. We begin by considering the biological significance of this. About 50 to 90% of the body weight of an insect, or nearly any other animal, is water. Some of this fluid is inside the various cells that make up the body while the remainder is in the spaces outside of the cells. As a unit, these spaces are regarded as a collective known as the extracellular space. The extracellular space is filled with fluid, for obvious reasons called the extracellular fluid. Among the dissolved constituents of the extracellular fluids are the ions and the nutrients needed by the cells for maintenance of life. The extracellular fluid is in constant motion throughout the body, and it is rapidly mixed to near homogeneity by hemolymph circulation and by diffusion. Because of these mixing processes, we can say that all cells in the body live in essentially the same environment. The great French physiologist, Bernard, called the extracellular fluid the melieu int'erieur or what we would call the internal environment.

The point is to recognize two main fluid compartments in an insect body. One is the intracellular compartment and the other is the extracellular compartment. To appropriate a teaching device from Daniel Matzia, the famed cell biologist retired from UC Berkeley, we can regard the extracellular fluids as the *out environment*, that is, the environment inside the body that cells experience. The intracellular environment, then, could be the *in* environment. However, we know eucaryotic cells have organells and compartments within them. Within cells, the compartment inside any particular organelle, such as a nucleus or mitochondrion, could be the *in-in environment*. At the organismal level, the world we live in is the *out-out environment*. This language allows us to appreciate the world in terms of environments that things experience: the in-in, the in, the out and the out-out environments. Cells experience the out environment, the same thing that Bernard called the *melieu int'erieur*.

Some percentage of the total body fluids of any organism are intracellular fluids. The fluid within each cell is made up of its own individual mixture of various constituents, such as ions, amino acids, proteins, sugars and so forth. On the other hand, within an organism, the concentrations of these things appear to be reasonably similar from cell to cell. For our discussion, the intracellular fluids of all of the different cells of an organism can be considered to be one large fluid compartment: the in environment or the intracellular compartment. Regardless of the particular language or expression that is easiest to keep in mind, we want to come to the point of view that an insect body has two major fluid compartments: the intracellular and extracellular compartments. The extracellular compartment is easy to visualize as the circulating hemolymph.

While the extracellular environment is substantially different in composition from the intracellular environment, the two compartments are in osmotic balance. That is to say that there is rarely a net flow of water from one compartment to another. There are various mechanisms to handle the maintenance of the correct concentrations of the intracellular constituents, including active transport and facilitated diffusion. These mechanisms are a related, but different matter. The thrust for our purposes is that water always tends to diffuse by a process called osmosis across cellular membranes from the side with fewer total dissolved particles to the side with a greater number of dissolved particles.

Let us consider some possibilities. First, if a cell in placed in a solution with far fewer osmotic particles than the cell, water will diffuse into the cell. This can result in considerable swelling, even to the point of bursting the cell. Alternatively, if a cell is placed into a solution with a far greater number of particles, water tends to diffuse, once again, down its osmotic gradient, in this case, out of the cell. A badly shrunken cell that may or may not survive may result from this flow. We gather the significance of a physical phenomenon, osmosis. Osmosis explains the importance of maintaining a proper balance of water and of salts in the extracellular compartment or the melieu int'erieur. During times when, for whatever reason, the extracellular compartment is not correctly regulated, the physiological situation is exactly like placing cells into a solution that is too dilute or too concentrated. This is a common problem faced by insects and most other organisms living in freshwater, salt water or in terrestrial environments.

Again, terrestrial insects are faced with the problem of regulating their levels of body fluids and ionic concentrations such that the extracellular and intracellular fluid compartments are maintained in osmotic balance. Failure in this respect would lead to this situation: the body's cells are effectively placed in a fluid that would cause them to swell or shrink to a point of dysfunction or death. Physiological solutions to the problem of water balance have tremendous evolutionary and ecological significance. First, it can be argued that the spectacular evolutionary radiation of insects occurred after a host of adaptations to terrestrial life. We saw some of these adaptations during our consideration of movement. Physiological regulation of water balance is another adaptation that supports terrestrial life. The biogeographical distribution of insects reflects the broad range of terrestrial habitats in which insects live and thrive. Many of these habitats are extremely arid and very hot. Physiological adaptations to these sorts of environments have permitted exploitation of such habitats that would otherwise be unavailable to insects. The focus on terrestrial life is not meant to overlook aquatic insects, which we will discuss in future. Furthermore, aquatic insects arose from their terrestrial counterparts. Hence, adaptations to aquatic life can be thought of as a component of successful terrestrialization. Two main factors may work to upset the essential osmotic equilibrium within an insect's body:

1) Loss of water by transpiration across the cuticle, or a very large gain of water by drinking or eating.
2) The intake of food whose concentration of inorganic salts is quite different from the ionic composition of the hemolymph.

Today we are going to begin with the first point, the regulation of water balance, which may be insulted by substantial loss or gain of water. Insects are generally small creatures, which tends to put them in an unfavourable setting with respect to the ratio of body surface area to body volume. The idea is to recognize a relatively large surface across which water can be lost by transpiration. We will see that water loss across the body surface can not be entirely stopped; however, if it were not restricted to a level that can be balanced by water uptake, insects would not survive in the more arid environments. The first line of defence was the development of a cuticle relatively impermeable to water.

The outer layer of an insect is the integument, which is made up of the epidermis and the cuticle. The epidermis is the outer cell layer. The epidermis lies on a basement membrante, and secretes the cuticle, which so far as I know is illustrated in virtually every book written on insect biology. There are three major sections of cuticle, endocuticle, exocuticle and epicuticle. In standard drawings, we note the pore canal, cells of the epidermis and the basement membrane. The endocuticle provides the extensibility of the integument and combines toughness with flexibility. The rigidity of the hard parts of an insect, the appendages and so on, comes from the properties of the exocuticle. The relative impermeability of the cuticle to water is due to the presence of the epicuticle. The epicuticle is a complex structure that differs among various insects.

There is a cuticulin layer made of two components: a thin, dense outer layer and a thicker, still denser inner layer. Cuticulin is generally made of lipoprotein. A waxy layer lies atop the cuticulin. This is made of long-chain hydrocarbons, and esters of fatty acids and alcohols. There is an inner wax layer composed of strongly oriented long-chain molecules that generally form a monolayer only one molecule thick. The strong orientation provides the waterproof properties of the epicuticle. The outer wax layer is composed of wax molecules that are randomly orientated; on top of this is the cement layer. People think that this is a shellac-like substance that may serve to protect the wax.

Terrestrial insects lose water by evaporation from the general body surface-a process called transpiration. The rate of water loss from the surface on any given insect is a function of both wind speed and also of the humidity of air. Transpiration is relatively independent of body temperature, up to some point. Water is lost from the body surface at a constant, but rather low, rate through a broad range of temperatures. At some temperature the cuticle reaches a temperature point called the transition point. This point varies from insect to insect, depending upon the chemical composition of the hydrocarbons and waxes in the epicuticle. Insects living in hot, arid habitats tend to have wax layers with higher transition temperatures than do insects living in cooler or more humid habitats. Some insects have two transitions points.

Water loss from the body surface is held to a low rate by the oriented wax layer on the outside of the cuticulin. When the temperature of the cuticle rises to such a point that the van der Waals forces holding the wax molecules together are broken, the molecules move slightly apart and water is allowed to escape. Below the transition temperatures the wax molecules are close together, and at a slight angle so that water loss is restricted. At temperatures above the the transition point, the molecules part, which usually leads to a quick and probably pleasing and comfortable death due to dehydration.

Then we see that osmotic balance in the melieu int'erieur is in part maintained by reduction of water loss from the body surface. Another important avenue of water loss in terrestrial insects is from the respiratory surfaces. In the orthopteran *Gastrimargus*, for example, 70% of the normal water loss is from the trachael system. We have already seen that water loss from the respiratory surfaces is restricted by the invagination, or internalization, of the trachael system such that it is inside the body. Water loss is also reduced by control of the spiracular valves. Internalization of the respiratory surfaces in a major adaptation to terrestrial life in vertebrates as well as in invertebrates.

We can model an experiment designed to demonstrate the influence of spiracles on water loss. Consider the rate of water loss from *Tenebrio* larvae as a function of time. *Tenebrio* is well adapted to arid conditions, and its water loss rates are typically quite low.

If we impose the condition of open spiracles, by holding the animals in an atmosphere of 5% carbon dioxide, water loss would be far higher than what could be easily balanced by an insect living in an arid habitat. Physiological regulation of the spiracles plays an important part in reducing potential water loss.

Some insects live through long periods of quiescence during which they do not feed or drink water. Control of the spiracles in these species is aimed at keeping them closed as much as possible to minimize water loss while facilitating the gas exchange necessary to support metabolic needs. One strategy is called cyclic release of carbon dioxide. This has been well studied in pupae of *Hyalopohora cecropia*, a large insect suitable for study in this regard. Carbon dioxide is allowed to build up in the trachael system for a long as seven hours, then is released in bursts during which the spiracles are wide open. Between the bursts of carbon dioxide release the spiracles are not closed as tight as possible, but rather flutter so as to allow a continuous uptake of oxygen.

While the cuticle and control of the spiracles serve to restrict the loss of water, a certain amount of water is inevitably lost. Then we can identify three avenues of water loss:

1) Transpiration across the cuticular surfaces
2) Transpiration across respiratory surfaces
3) Loss of water in the excreta.

Water that is lost must be balanced by an appropriate level of water uptake. Most insects take in sufficient water with the food they eat. In his book on insect structures and function, Chapman thought some insects may select food with a higher water content under certain circumstances. Still other insect species may consume more food than they actually need in order to realize the potential water in the food. Moreover, many insects drink water. The dipteran *Phormia*, for example, has specific water receptors in the tarsal and labellar sensilla. I've often observed wasps drinking water in desert pools in Mexico.

Some insects have evolved slightly more bizarre adaptations to gain water. Larvae of the dipteran, *Epistrophe,* can extrude an anal pipilla into a water droplet and absorb water. The cockroach *Periplaneta* is among many insects that can absorb water from water droplets on the cuticle. The cuticle is asymmetrical with respect to water movement: water passes inwards more readily than outwards. Also, a few insects can absorb water from water vapour in the air when they are desiccated. *Tenebrio*, for example, can absorb water from air with relative humidities greater than 90%. Prepupa of the flea *Xinopsylla* can absorb water from air at relative humidity above 50%.

A final source of water is called metabolic water. As we saw during our look at metabolism, the final electron acceptor at the end of the electron transport chain is oxygen. Electrons and their hydrogens combine with oxygen at this point, yielding water. The amount of water that can be gained by metabolism depends on the metabolic substrate. A hundred grams of the fatty acid palmitate (16:0) yields 112 grams of water while 100 grams of glycogen yields 56 grams of water. Grain insects, such as *Tribolium*, normally obtain a very large proportion if not all of their water from metabolism of their food. We can guess that metabolism of lipid may be adaptive with respect to water balance during long periods of hibernation in several insect species.

Osmoregulation in Terrestrial Insects: Salt Balance

We can regard the current line of talks as a discussion of physiological adaptations to terrestrialization. Perhaps you can try to integrate several areas of physiology along those lines, including information about the integument, the respiratory system and the circulatory system. In our last meeting, we looked at the business of maintaining an optimal osmotic balance between the two major fluid compartments in insects, the intracellular compartment and the extracellular compartment. We saw that the osmotic balance was, in part, achieved by mechanisms that restrict and offset the unavoidable loss of water that must necessarily influence terrestrial organisms. I emphasize the water loss is unavoidable. In particular, we saw that the properties of the cuticle, control of spiracular openings and resorption of water from faeces are important mechanisms to control water loss. There are also a variety of mechanisms to replenish water supplies.

Today we want to continue to develop our appreciation of osmoregulation by observing that organisms must an optimal balance of water and of salt within the body. We develop the problem by considering some specific data. A brief look at the concentrations of various ions in an insect's regular food and in its hemolymph shows a couple of points. First, the compositions of the the food and of the hemolymph are remarkably different in many cases. Second, and more important, the differences are not uniform across the range of individual ions.

Here are examples. Larvae of the bot fly *Gastrophilus* are internal parasites of horses, mules and some larger wild animals. In this insect sodium is much higher in the hemolymph than in the food, while potassium is higher in the food than in the hemolymph. A different situation obtains in the stick insect *Dixippus*, which often eats Privet leaves. Privet comes from shrubs in the olive family. Here, sodium, potassium and calcium are much higher in the food while magnesium is much lower. Finally, in the Colorado potato beetle *Leptinotarsa*, there is very little sodium in the common food while potassium and calcium are much higher in the food than in the hemolymph. This table presents the specific data:

Cation composition of hemolymph and food, given in milliequivelents/liter.

Insect species	–	–	–	
Food	Sodium	Potassium	Calcium	Magnesium
Gastrophilus	175	12	6	32
Horse blood	85	31	2	3
Dixippus	9	28	16	142
Privet leaves	46	152	825	40
Leptinotarsa	4	65	48	189
Potato leaves	--	145	129	86

Now, if these salts were not somehow regulated, an insect would face the problem of osmotic, rather than water, imbalance between the two major fluid compartments every time it takes a meal. A clear example of this would be the large change in hemolymph calcium ions following a typical meal in the stick insect. So we want to ask: How are hemolymph ion concentrations regulated in the face of regular insult with each meal? Just imagine.

One mechanism to regulate ion concentrations occurs at the level of the alimentary canal. The insect would simply limit the absorption of unnecessary salts. They would absorb only the required organic food material while unnecessary inorganics were passed though the insect and voided in due course. However, that is not the way hemolymph ion concentrations are regulated. In point of fact, dietary inorganic salts are readily absorbed across the epithelium of the alimentary canal and the ions therefore mix with the general pool of inorganic material in the extracellular compartment. The physiological problem, then, becomes one of moving ions into and out of the extracellular compartment in such as way as to sustain an optimal osmotic balance.

The most general expression is that secretion occurs in the Malpighian tubules and resorption takes place in the rectum. A generalized model of the alimentary canal shows the foregut, midgut and hindgut. Within these familiar major regions are the Malpighian tubules, which most often join the alimentary tract in the midgut-hindgut junction, and the rectum, an enlarged portion of the hindgut featuring specialized epithelial cells. We want to appreciate the various sorts of movements of materials across the different regions of the alimentary tract. This is to be taken rather as a composite, and represents all sorts of movements, some of which do not occur in all insects. In due course we will see that water, inorganic ions, sugars and amino acids are mostly absorbed in the midgut.

With respect to ion concentrations, we want to see that water and ions are secreted into the lumen of the Malpighian tubules. These components are moved into the hindgut by mass flow, and then on to the rectum. Depending upon the individual needs of the insect, water and ions are reabsorbed by cells in the rectum and moved back into the hemolymph. We might get the impression that the same hemolymph constituents are moved in a sort of futile cycle, but we will see that the situation is otherwise.

Malpighian tubules are closed-ended tubules that arise from either the posterior part of the midgut or the anterior portion of the hindgut. The precise anatomical arrangement differs among insect groups. The walls of the tubules are one cell layer thick. The cells comprising the tubule walls stand on a basement membrane. The tubules from most insects are surrounded by muscles which move the tubules around in the hemocoel, and the tubules are supplied with numerous tracheae. Malpighian tubules of mosquitoes and some other insect species do not have muscles.

There are distinct histological and functional regions within the overall structures of Malpighian tubules from many, but not all, insects. The tubules have proximal and distal portions, proximal being closer to the attachment with the alimentary canal. The distal portion is functionally the secretory region, and features cells with honeycomb borders. By secretory we mean that water and ions are secreted from the hemolymph into the lumens of the tubules. The proximal region is composed of cells with brush borders; this is the absorptive region of Malpighian tubules.

We want to appreciate the cellular structure. The cells attach to the basement. Cells from the distal region feature a honeycomb border formed by cytoplasmic filaments. These filaments are rich in mitochondria, which supply the energy required for active transport processes. The cell membranes on the basal side are envaginated. All of these features are consistent with the secretory function of these distal cells. Cells from the proximal region of the tubule are different from the cells of the distal region in many insects. These cells feature a brush border. They are made up of more widely separated microvilli and have less complex envaginations of the basal-side membranes. They have more mitochondria in the basal side of the cell, thought to support their reabsorptive role in the tubule. Again, not all insects have Malpighian tubules of this heterogeneity. We also note that some insects, such as aphids, collembola and thrips have no Malpighian tubules or only small absorptive pads called papillae.

The Malpighian tubules of some insects have the distal ends adpressed to the sides of the rectum. The ends are surrounded by a membrane known as the perinephric membrane. The perinephric membrane forms a cavity between the rectum and the general hemolymph. This is known as a cryptonephretic arrangement. In the meal worm beetle *Tenebrio* this is thought to help absorb a maximum amount of water from feces.

A primary urine that is isosmotic to the hemolymph is formed in the Malpighian tubules. A general physiological observation is that water is rarely moved across membranes by itself. One way to remember this is to recall that ions can be moved by active pumping processes and water follows the locally formed gradient. In Malpighian tubules, fluid secretion is driven by active transport of potassium and chloride. In some blood feeders sodium is also actively transported. Potassium chloride (KCl) is usually the major solute in the lumenal fluid, and the concentrations of the ions are usually low compared to the hemolymph. The Malpighian tubule fluid makes its way to the alimentary canal by bulk movement aided by muscles. The fluid then moves through the hindgut to the rectum. In the rectum selective reabsorption of water, ions and various metabolites occurs by active transport processes. It is only in the rectum that the urine becomes strongly different in osmotic concentration from the hemolymph. Both tubular secretion and rectal absorption are under separate endocrine controls which together determine the amount and ionic composition of water that is excreted.

The secretion and absorption processes are summarized in the figure. This drawing is derived from the blood-sucking bug Rhodnius. This represents those insects with heterogeneous Malpighian tubules. In the distal region of the tubules potassium, chloride and sodium are pumped into the lumen by active processes; water follows the ion gradient into the lumen. Note that the figure does not show the movement of chloride. We will talk about uric acid formation and excretion later, but for now please note that uric acid also moves into the tubule. Resorption occurs in the proximal portion of the tubule, with potassium moved by active processes, and water following. After movement of urine and uric acid crystals to the rectum, more water can be resorbed, along with selected ions.

Some insects have homogeneous Malpighian tubules, such as the stick insect also represented in the figure. Secretion takes place all along the homogenous Malpighian tubules. Some movements are active processes, such as sodium and potassium pumps, while others are passive. There is no resorption until urine is moved into the rectum, where several ions, including amino acids are moved back into the hemolymph by active processes.

Again, here are data to make the point. This information comes from work on the locust, all in meq/liter:

Osmotic pressure and concentrations of selected ions.

Compartment	Osmotic pressure	Sodium	Potassium	Chloride
Hemolymph	214	108	11	115
Primary urine	226	20	139	93
Rectal fluid	433	1	22	5

These data show specific reabsorption of ions on an individual basis. The hemolymph is typically high in sodium, while potassium is the major urine ion in the tubules. These data indicate that sodium and chloride are selectively resorbed in the rectum. The osmotic concentration of the hemolymph and the urine are about the same, although of differing ionic compositions. The fluid components of the rectal contents are much more concentrated than the other fluids. These data come from locusts fed with fresh water. By way of helping to prepare for any future exams that could crop up right out of a clear blue sky, consider what these data might look like if the locust had been fed salt water. There would be high sodium levels in the rectum, as an avenue to excrete excess sodium. Again, ions are individually regulated and transported. The key point is that the Malpighian tubules secrete water and salts in urine formation. Superimposed upon this urine formation mechanism is a resorption mechanism to facilitate the selective uptake from the urine of those specific things required to sustain the optimal osmotic balance between the intracellular and extracellular fluid compartments.

Consider these situations that an insect may confront::

- subjected to greater than normal desiccation : this situation may result in increased hemolymph osmotic pressure. The insect may respond with decreased urine formation, and greater rectal water uptake.
- subjected to greater than normal desiccation and a food higher than normal in salt : this situation may result in increased hemolymph osmotic pressure, but the increased salt would not allow a maximal decrease in urine production. The insect may respond with a greater uptake of rectal water and a decrease in salt resorption.

From a biological point of view, Malpighian tubules in some insects are adaptive in unusual functions. Larvae of the face fly *Musca autumnalis* have four Malpighian tubules, two anterior and two posterior tubules. Both pairs arise from a single junction between the mid-and hind-guts. The posterior set is found in close association with the midgut. The anterior set expands distally, and is filled with a granulated liquid.

The posterior Malpighian tubules are regarded as ordinary secretory organs. The anterior tubules play a special role in the physiology of pupation. The puparium of this fly is hardened by deposition of calcium salts, rather than by the usual phenolic cross-linking typical of other insects. Dietary calcium salts, as well as other minerals, are stored in the anterior Malpighian tubules during the larval stages. At pupation, the calcium salts become dissolved, then transported from the tubules into the hemolymph. The calcium is directly transported from the hemolymph to the cuticle.

The Malpighian tubules of lacewing larvae undergo morphological changes before pupation. The modified tubules then produce silk for the pupal stage.

Physiology of Diapause

During our last meeting we considered one aspect of living in temperate and polar regions of our planet, that is the business of cold hardiness in insects. We emphasized mechanisms of resisting and tolerating freezing. Biosynthesis of antifreeze compounds allowed body fluids to cool far below a normal freezing point without ice crystal formation. This process is known as supercooling. In addition to supercooling, certain insects can tolerate freezing, for which the most important adaptation is that they can tolerate a remarkable level of cellular dehydration. These physiological adaptations to life in very cold conditions can be taken as rather extreme examples of a sort of biological recognition that conditions in almost every environment, aquatic, marine, terrestrial or atmospheric, are subject to regular and irregular changes. Today we want to regard some of the kinds of changes in environment, and look at some of the ways insects respond to changes.

Many insect species adapt to regular, seasonal changes in environmental conditions by manifesting various expressions of diapause. We want to appreciate what diapause is, and touch upon some aspects of physiological regulation of diapause. Tauber developed a comprehensive definition of diapause: *Diapause is a neurohomonally mediated, dynamic state of low metabolic activity. Associated with this are reduced morphogenesis, increased resistance to environmental extremes, and altered or reduced behavioural activity. Diapause occurs during a genetically determined stage of metamorphosis, and its full expressions develops in a species-specific manner, usually in response to a number of environmental stimuli that precede unfavourable conditions. Once diapause has begun, metabolic activity is suppressed even if conditions favourable for development prevail.*

This is a large definition, and it is meant to convey several points: First, diapause is not a single physiological event, but rather is a syndrome of several physiological and behavioural manifestations all bearing on adapting insects to forthcoming environmental changes. Second, the environmental signals that are translated into a diapause syndrome are called token stimuli because they do not themselves impact on insect life. Token stimuli simply herald the coming changes.

Third, the definition is biologically large because it includes dormancy and seasonal migration and seasonal polyphenism. To keep our terms straight, let us agree that dormancy is a period in a life cycle during which growth, maturation and reproduction are held in check. Dormancy can occur in any season. Polyphenism refers to seasonal changes in colour or structure of an organism. There are many examples of this among insects, including increased wax secretion, variations in wing length and quality of cocoons (tighly versus loosely spun).

Fourth, the events in a dispause syndrome are not aimed directly at organismal survival during any particular set of environmental conditions. Diapause has more to do with synchronization of entire life cycles to seasonal changes in environments. It affects survival during unfavourable conditions, but it also bears a large weight upon other features of life history, including growth, development and reproduction. Hence, diapause is a unifying biological concept. Because it is under genetic control, it is an important substrate in evolution of life histories and in speciation.

Insects also cope with unpredictable environmental changes, such as droughts and overpopulation. Short-term or localized changes might include sudden temperature changes or local food shortages. Insects can adjust to these sorts of changes with aseasonal quiescence or aseasonal migration. These adaptations differ from diapause in the manner of physiological control. We will see that diapause is controlled by endocrine regulation; aseasonal adaptations are under central nervous system control.

Diapause is a genetical feature of insect life histories. All metamorphic stages in life cycles can be utilized as a diapause stage, but diapause usually occurs in a stage characteristic of a given species. Recalling from the definition that diapause is a phenomenon that anticipates future seasonal changes, we see that token stimuli must be perceived by an insect, then translated into a biological response. These token stimuli are perceived only during specific stages, and as with the diapause itself, timing of perception of token stimuli is under genetic control.

Perception of token stimuli and expression of diapause can be rather close in time, or widely separated between two generations. Here are some examples of the possibilities. Many insect species undergo diapause during the egg stage. In the cricket *Teleogryllus*, young eggs are sensitive to diapause-inducing stimuli, and diapause follows in eggs that are slightly further along in development. There are also cases in in which diapause-inducing stimuli are perceived during various stages in development of the mothers. Finally, in the aphid *Megoura* the grandparental generation is the sensitive one.

Let us consider diapause in context of overall seasonal cycles. First, diapause has recognizable stages. The *prediapause period* is regarded as a time of feeding and metabolic reserve accumulation, as well as growth to the species-specific diapause stage. The *induction period* is usually characterized by reduction in feeding, metabolism and growth. The *maintenance period* sets the length of diapause, and is thought to be genetically regulated. During this time insects are sensitive to token stimuli, which can affect the intensity and length of diapause. These two remarks are not to be taken as contradictory, but rather as an indication of a complex regulatory system. Diapause is broken by token stimuli, which entrains a series of physiological events leading to increased metabolism and post-diapause transition to non-diapausing condition.

Going back to the idea of a diapause syndrome, we want to emphasize that diapause is not simply a reduction in metabolism and growth, but is a combined expression of several related, but different, physiological adaptations. These can include prediapause behaviours (such as preparing a hibernaculum), accumulation of glycerol for cold-hardiness, and colour changes to track changes in plant colours, as well as alterations in energy metabolism. Insect endocrine systems play a central role in all of these events. At one time it was thought that diapause was mediated by a single endocrine mechanism for each life stage, but, as usual, the situation is more complicated.

It is generally agreed that diapause in adult insects attends reduced corpora allata activity, and hence, reduced hemolymph concentrations of juvenile hormone (JH). As the name suggests, we usually regard JH in terms of insect developmental physiology. However, its actions in regulating developmental events in juvenile insects is just one aspect of JH biology. JH is normally present in adult insects. JH is involved in regulating aspects of sexual maturation, such as egg development in females of some species and accessory gland development in males of some species. It follows that reduced JH titres (in endocrinology, titre refers to circulating levels of a hormone) would at least impact on reproductive aspects of diapause. In some cases adult diapause can be terminated by treating diapausing adults with JH.

Adult diapause is induced by short day lengths in the Colorado potato beetle. Diapause can be induced in beetles maintained under long day conditions by extirpating the corpora allata. In this species corpora allata activity is regulated by stimulatory neurohormones from the brain. Termination of diapause, accordingly, requires long day conditions.

The sources of the stimulatory neurohormones are elements of the pars intercerebralis and corpora allata. Hence, in the Colorado potato beetle complete regulation of diapause involves neurohormones and JH. Just to give you a more detailed picture, the case is more complicated still in this beetle. Individual features of the diapause syndrome appear to be regulated by specific titres of JH, so that JH probably plays several different regulatory roles.

Diapause is under slightly different regulatory mechanisms in the bug *Pyrrhocoris*. Here JH and neurohormones are involved, as in the potato beetle. This bug differs, though, because there are inhibitory neurons from the brain that directly inhibit corpora allata activity, and thereby maintain low JH titre and diapause. The point here is recognizing an element of direct neurological control of diapause in this insect.

The classic work on pupal diapause was carried out on the giant silkworm, *Hyalophora cecropia*. When given their choice, physiologists often work on larger insects. The work on the giant silkworm was carried out by Carrol Williams and his co-workers at Harvard University. I guess you might say at the Nebraska University of the east coast. In this species pupal diapause follows cessation of brain neurosecretory activity, which stops prothoracicotropic hormone (PTTH) release. When we treat the physiology of post-embryonic development we will see PTTH stimulates prothoracic glands to release ecdysteroid, the moulting hormone. For our current interest in diapause, we note after a period in cold temperatures, "reactivated" brains release PTTH. PTTH stimulates release of moulting hormone, and this hormone signal leads to active adults.

The pupal diapause of another lepidopteran, *Heliothis zea*, is independent of PTTH. Low temperatures somehow act directly upon prothoracic glands, which reduces titres of moulting hormone. Pupae of flesh flies break diapause in response to moulting hormone, but the precise mechanism is slightly different from the lepidopterans just discussed. In flesh flies a sharp but brief peak of JH occurs before release of ecdysteroid. The JH is thought to positively influence, or somehow prepare, tissues to respond to ecdysteroid.

Larvae of many insects undergo diapause. In some speces, induction and maintenance of diapause depend on continuous presence of JH. In the rice stem borer, *Chilo suppressalis,* JH titres remain high throughout diapause, and corpora allata activity decreases as diapause draws to an end. This physiological mechanism of regulating diapause occurs in other species, too, but does not explain diapause in all insect larvae. In another mechanism, some diapausing larvae continually secrete JH from active corpora allata. In these species, diapause is induced and maintained because the JH inhibits release of PTTH.

Diapause interupts embryological development at a very early stage in the eggs of some insect species, such as the silkmoth *Bombyx mori*. This results from a maternal diapause hormone produced by the mother. Maternal diapause hormone is produced in the subesophageal ganglia, then transported via hemolymph circulation to eggs within the ovary. Circulating maternal diapause hormone causes diapause in eggs. A similar mechanism may occur in other insects, including the bug *Adelphocoris*.

Maternal diapause hormone is released from the subesophageal ganglia. The hormone release is triggered by enviornmental events. Silkworm females which experienced long days and warm temperatures during their egg and early larval periods lay eggs that enter diapause. Females which experienced short days do not produce maternal diapause hormone, and their eggs do not enter diapause. The maternal diapause hormone of *Bombyx* has been purified. It is a dipeptide, consisting of two different peptide chains.

Diapause hormone acts to arrest embryonic development at the stage of germ band formation. There is a tight correlation between diapause and carbohydrte metabolism. The developing ovaries accumulate large amounts of glycogen during the induction phase of diapause. When diapause begins, the glycogen is converted to polyols. At the end of diapause the polyols are converted back into glycogen, which becomes the major source of energy to carry out embryonic development. Diapause hormone is thought to influence carbohydrate metabolism in eggs.

The follicle cells of the ovaries are the targets of diapause hormone. Let us consider a model of diapause hormone action. Diapause hormone is thought to interact with specific receptors on the plasma-membrane of follicle cells.

The hormone receptor translates the hormone signal into reduced intracellular levels of cGMP. The decreased cGMP levels activates increased membrane trehalase activity. This activity leads to increased trehalose hydrolysis and in-bound transport of glucose. The glucose is rapidly taken up into glycogen. Beside its effect on carbohydrate metabolism, diapause hormone is thought to exert other actions on ovaries.

The physiology, ecology and evolution of insect diapause remain contemporary areas of study. A far more complex picture of hormonal regulation of diapause will emerge in due course.

Respiration in Aquatic Insects

One of the major issues in the physiological ecology of aquatic aerobic animals is the availability of oxygen. Compared to atmospheric chemists and other people along those lines, physiologists take a rather simple view of air. We generally consider air is composed of something like this:

Element	Proportion
Oxygen	21%
Nitrogen	79%
Carbon dioxide	0.03%

We can quickly gather a couple of points from this information. First, there are several trace elements in air that are lumped into nitrogen for purposes of discussion. This is acceptable because the quantitatively minor gasses, such as argon, are also inert, and physiologically not different from nitrogen. The second point is that we can note from the percentage composition that a litter of dry air at sea level contains about 210 ml of oxygen. Just from watching animals moving around on land, 210 ml of oxygen per litter of air seems to be plenty of oxygen.

Well, Ok, let us ask how much oxygen is in water? The answer is not quite so simple, even for physiologists. To correctly look at availability of oxygen to animals living in water, we need to consider several points.

First, gases are not equally soluble in water. Hence, the solubility of individual gases will partly determine availability of oxygen. Here are a few data on solubility of pure gases in distilled water at 15 oC and 760 mm Hg (1 atmosphere of pressure):

Element	Solubility
Oxygen	34.1 ml/liter
Nitrogen	16.9 ml/liter
Carbon dioxide	1019.0 ml/liter

We can use these solubility data to calculate volumes of gas available in water when the water is in equilibrium with air at one atmosphere:

Volume of gas = SOLUBILITY × % COMPOSITION

Element	Solubility × % composition	Volume
Oxygen	34.1 × 21%	7.16 ml/liter
Nitrogen	16.9 × 79%	13.35 ml/liter
Carbon dioxide	1019 × 0.03%	0.31 ml/liter

Now with just a little bit of high-powered mathematics, such as long division, we can calculate that the 7.16 ml/l of oxygen in water is about 30 times less than the 210 ml/l of oxygen in air. We can also calculate the percentage composition of gases in water at 15 oC and one atmosphere pressure by adding the three volume values and dividing each by the total:

Element	*Volume/total*	*Percentage composition in water*
Oxygen	7.16/20.82	34%
Nitrogen	13.35/20.82	64%
Carbon dioxide	0.31/20.82	0.01%

These values hold for water at 15 °C, equilibrated with air at one atmosphere, typically regarded as sea level. The amount of gas in water, however, varies with water temperature, among other physical parameters. The following table will make the point:

Water temperature (°C)	*Volume of oxygen (ml/liter)*
0	10.3
10	8.0
15	7.2
20	6.6
30	5.6

This table shows an inverse relationship between water temperature and oxygen contents, that is to say, there is a constraint on oxygen availablity in water due to the influence of temperature. This temperature constraint presents a problem because most aquatic animals, particularly small animals such as insects, are poikilothermic. The body temperatures of poikilothermic animals vary with ambient temperature: as water temperatures increase, so do body temperatures. The problem stems from the influence of temperature on metabolic rates. Metabolic rates are directly related to body temperatures. As metabolic rates increase, oxygen consumption rates increase. The environmental influence of water temperature is seen in body temperature, which influences metabolic rates. These factors intersect at a crucial, yet paradoxical point: increasing oxygen requirements are attended by decreasing availability of oxygen in water.

Respiration in aquatic animals, then, is a look at physiological solutions to the problem of obtaining sufficient oxygen to support metabolic needs from an environment where oxygen is in short supply.

From the perspective of environmental scientists, we note oxygen is one of the physical parameters of surface fresh water that change markedly in reaction to the various of the polluting chemicals regularly dumped into lakes, rivers and streams. Many aquatic insects and certain of our favourite fresh water fishes can not survive in waters with reduced oxygen levels. The presence or absence of these animals that require higher levels of disolved oxygen can be used to assess the relative environmental health of a water system. Such species are referred to as indicator organisms.

More germane to our topic, however, is understanding that insects and other aquatic animals have potential problems of low oxygen availability. One solution to the problem of obtaining oxygen while living in aquatic environments is to directly breathe atmospheric air. Many insects have respiratory siphons. In insects with this sort of anatomical arrangement only the posterior spiracles are functional, and they are located at the end of a siphon such that only the posterior portion of the insect penetrates the surface film of water, while the rest of the body is submerged, often suspended from the surface film. The respiratory siphon on larvae of drone flies, *Eristalis*, is telescopic, not meant to suggest one can see through it. Mosquito larvae also have respiratory siphons, but they are not so long as in drone flies.

One constraint on the function of respiratory siphons is that the tips of siphon tubes must be arranged such that air can move into the siphon while water is excluded. Commonly, siphons have hydrofuge hairs associated with their spiracles. When their spiracle is submerged the hairs close over the spiracle and water is kept out of the siphon. At the surface, the hairs are separated by surface tension forces, allowing air to pass to and fro within the siphons. There is typically a funnel shape at the surface.

Respiratory siphons have certain shortcomings. The siphons tend to limit insects to the upper levels, or strata, of water, so that they are close to the surface. Insects require other adaptations to achieve some independence from the surface strata. Some insects manage to exploit deeper reaches of water by carrying air stores along with them during their dives. This strategy is well developed in the predacious diving beetle *Dytiscus*. By keeping air stores in contact with spiracles during a dive, these beetles have their trachael systems plus air stores to draw upon to meet oxygen requirements.

Spiracles are typically located along the sides of many terrestrial insects, but this arrangement does not help much with carrying supplies of air under water. Many beetles have their abdominal spiracles on the dorsal surface, under the elytra. This needs be only slightly adjusted for carrying air stores during diving. Many other insects, especially aquatic true bugs also carry air stores under water.

The efficacy of the air store is vastly improved because the bubble itself can act as a physical gill. That is, it can extract oxygen directly from the surrounding water. The following work on the backswimer *Notonecta* illustrates the point. When this bug was held under water that had been equilibrated with pure nitrogen, it survived for only 5 minutes. When the water was equilibrated with air the insect could survive under water for 6 hours, 72 times as long. These two observations are consistent, and they can be understood with respect to extraction of oxygen from the water. The importance of nitrogen in the air bubble can be shown by the following experiment. If *Notonecta* is given a gas bubble of pure oxygen, instead of an air bubble, the bug can survive for about 35 minutes under water. The role of nitrogen in performance of a physical gill can be explained as follows:

Soon after a bug dives, a disequilibrium occurs. Removing oxygen from the bubble by metabolism decreases the partial pressure of oxygen compared to water and increases the partial pressure of nitrogen. Oxygen tends to move down its concentration gradient into the bubble while nitrogen moves down its gradient from the bubble into water. The solubilities of gases become important. Nitrogen has low solubility in water, and it diffuses out of the bubble more slowly than oxygen diffuses from the water into the bubble. The nitrogen is essential to the bubble as physical gill because without it there would be no change in partial pressure of oxygen; the differential partial pressures drive oxygen from the water into the bubble.

This system can provide enough oxygen to small, inactive insects to allow them to remain submerged for months. More active insects require more oxygen. The diving beetles have to surface every few minutes to replenish their oxygen stores. In addition, certain physical parameters assert themselves. For one, as we saw earlier, water temperature effects the amount of oxygen that can be dissolved in water. At higher temperatures there is less oxygen in water, and at the same time an insect requires, and consumes, more oxygen. A physical gill may be of small value under such circumstances. Another factor impacting on air bubble life spans is depth in water. For every meter of depth in water there is about a tenth of an atmosphere increase in pressure. Increased pressure speeds diffusion out of air bubbles, and shortens their useful life span. Here is a summary of several factors that impact on the useful life spans of air bubbles, and thereby determine how long an insect may survive under water:

1. Temperature of water
2. Amount of O_2 in water
3. Metabolic rate
4. Size of bubble
5. Depth of dive.

Now, when an insect dives with an air bubble, nitrogen slowly diffuses into the water, and the bubble becomes progressively smaller until it collapses. The collapse is delayed because nitrogen moves into water only slowly, but sooner or later, air bubbles will collapse. Insect using air bubbles as physical gills must return to the surface. If an air bubble were non-collapsible, many aquatic insects would not have to surface. Some insects, such as the water bug *Apohelocheirus* are equipped with a plastron, which holds a non-collapsible thin film of air along the outer surface of the body. A plastron is continuous with the trachael system. A plastron is a dense (up to 2.5 million hairs/ square mm of surface area) layer of stiff hydro-phobic hairs.

Plastrons typically run along the body surface, continuous with the spiracles. There are cuticular openings which lead into air channels, which in turn lead to spiracles and trachael systems. The hydrophobic quality of hairs along the air channels serve to keep water out of trachael systems.

The most efficient arrangement of plastron hairs would be to have them parallel to a body surface. Insects approximate this by having the tips of plastron hairs bent, or else they slope as some angle, so that air films resist collapse when subjected to higher pressures. Plastrons function in the same patterns as air bubbles when they are extracting oxygen from water, that is to say, they act as physical gills. Plastrons are found on surfaces of many insect eggs and on many species with aquatic pupal stages.

We have seen two approaches to organismal respiration in aquatic environments. One is to breathe air directly by way of siphons and another is to extract oxygen from water using a sort of physical gill. Several groups of insects express a third approach, which is to develop a true physiological gill to extract oxygen from water. The term gill, with no prefix such as "true" or "physiological" is sufficient because other sorts of gills are linked to an appropriate modifier, such as anal gill or physical gill. Gills are appendages for gas exchange, and tracheae branch into these appendages. Oxygen diffuses from water into tracheae of gills, and from there moves under the usual motivation by way of trachael systems to tissues.

The locations of gills varies in larvae of several insect species. Stone fly nymphs, *Pteronarcys,* have gills on the thorax and first two abdominal segments. Gill plates are located at the end of abdomens of damselfly larvae. Finally, mayfly nymph have gills on thoraces and abdomens. Just in passing, the gills of mayfly nymphs are equipped with nervous and muscular connections. The gills can be swept back and fro to mix water, and thereby alleviate the problem of local oxygen depletion. This arrangement is the source of the tracheal gill theory in arguments over the evolution of insect flight.

Gills extract oxygen from surrounding water. Due to its low diffusion rate in water, oxygen can be depleted from water immediately local to respiratory surfaces. This local oxygen-poor water must be replaced by fresh, oxygen-rich water. Moving water for this purpose is called ventilation of the gills. Some insect ventilate their gills by selecting a microhabitat where water is flowing. Still others ventilate by muscular action. This can require considerable energy expenditure because water has greater density and viscosity than does air. The oxygen costs of ventilation must be more than offset by oxygen gains. Work on caddis fly larvae illustrates the point. In the experiments, metabolic rates of caddis fly larvae increased when water current velocities decreased. These data suggest the larvae can somehow register changes in oxygen availability, then respond by ventilation. You can try it yourself. Hold your head underwater for some time, see if you don't register a change in oxygen availability.

Insects use a variety of methods to ventilate their gills. Mayfly nymphs beat their abdominal gills in the absence of water movement. Stonefly nymphs ventilate by doing sort of "push-ups" at a rate related to amounts of oxygen in water. Larval dragonflies combine ventilation of gill surfaces with locomotion. Dragonfly larvae aspirate water into their rectums, which feature rectal gills. These larvae can locomote by jetting themselves forward. They do this by rapidly squirting aspirated water back out of their rectum. Damselfly larvae use their abdominal gills as paddles to propel themselves while simultaneiously ventillating their gills.

In vertebrate respiratory systems the respiratory pigment hemoglobin serves three functions. First, it withdraws oxygen from the air at respiratory surfaces in lungs. Second, it transport oxygen throughout bodies. Third, hemoglobin releases oxygen to tissues at controlled rates. Some insects also contain hemoglobin molecules, but the role of insect hemoglobin is storage, rather than transportation, of oxygen. Insect hemoglobin is about half the molecular weight of vertebrate hemoglobin, and it has a much higher affinity for oxygen. That is, it can be saturated at much lower oxygen tensions. Finally, insect hemoglobin is disolved in hemolymph; it is not carried in blood cells. The larvae of certain midges living in anoxic mud have hemoglobin in their hemolymph. To emphasize the point, these larvae tend to look sort of pink when they are well oxygenated. They live in burrows, in which they periodically wiggle, setting up currents of water. The fresh water contains oxygen, which is taken up by the insect hemoglobin. The hemoglobin can store enough oxygen to support about 9 minutes of metabolism.

Two aquatic hemipterans, *Buenoa* and *Anisops*, also store use hemoglobin to store oxygen. Part of their oxygen stores for dives is in a small bubble, the rest is in hemoglobin. Their bubbles are small enough the insect can have neutral buoyancy and remain suspended in water while awaiting prey.

It has been suggested that with larger bubbles they would not be able to suspend themselves in water, and without hemoglobin they would have to surface much more often. In this case, the hemoglobin seems to be an adaptation for predation in water.

Finally, another adaptation for obtaining oxygen while remaining submerged is to extract oxygen from aquatic plants. Some insects do this by thrusting their spiracles into the aerenchyma of plants; their spiracles are located at the tip of a sharp-pointed post-abdominal siphon. Larvae of two dipterans, *Chrysogaster* and *Notiphila* do this, as well as larvae of the beetle *Donacia*. These insects live in oxygen-poor mud.

Insect Immunology: Humoral Immunity

One question entomologists and university administrators might wonder about now and again would be along the lines of "What are some of the emerging areas of insect physiology that are likely to have great impact on how entomology might be carried out in the future? Several topics come to mind, including understanding the roles of peptide hormones, the significance of biogenic amines, more detailed appreciation of sensory physiology, and intracellular signal transduction mechanisms. However, above all of these topics, and this is certainly not to be taken as an exhaustive list, the study of insect immunology may well develop into one of the most active and most important areas of insect physiology.

Insect Diseases

The idea that insects can have diseases goes well back to antiquity. The idea predates the science of microbiology by centuries. The earliest records come from descriptions of a pest of honey bees by Aristotle. In the infancy of microbiology, Pasteur showed that two diseases of silkworms were caused by microbes. As you might expect, much of Pasteur's work was motivated by economic concern: the silk industry suffered considerable losses caused by diseased silkworms.

In more recent times insect pathogens have used microbes as agents of biological control of insects. Although many microorganisms have been studied as actually or potentially useful control agents, none have so far received as much attention as *Bacillus thuringeinsis*. The potential of various microbial forms, including bacteria, fungi and protistans, as insect control agents has driven an impressive body of work on insect microbiology and pathology.

The study of insect immunity went along with much of the very early work on the general question immunology. There was a period of considerable activity in the 1920's. A second surge of work on insect immunity started in the late 1950's, when it was discovered that a certain immunity could be induced in some insects. This induced immunity can be compared to a sort of standing, or constitutive, immunity. The most recent period of research on insect immunity began in 1980 with the first isolation and purification of an induced antibacterial factor. Modern work on insect immunity is particularly satisfactory because it is one of the areas of insect biology where molecular biological tools have yielded a rich harvest of information that would not otherwise have been generated.

It is appropriate to define immunity. Immunity is "resistance to or protection against a specified disease". This definition draws attention to the phenomonology, and it is independent of specific immunological mechanisms. The distinction is important because the contemporary wisdom in some circles holds that insects do not have an immune system because they do not produce antibodies of the mammalian sort.

The Major Immunity Mechanisms

There are two major areas of insect immunity, cell-free immunity and hemocytic immunity. These are sometimes called humoral immunity and cellular immunity, respectively. These two immunological responses are complimentary, and both may be seen in response to the same infection. Please note that immunity in insects and in other invertebrates differs from the immune systems of mammals in fundamental ways. First, insects produce no lymphocytes, of which there are many sub-types including eosinophils and neutrophils, no T-cells and no helper cells. So the cellular immunity of insects is not to be compared to the cellular immunity of mammals, although some people have tried to make that sort of comparison. Second, insects do not biosynthesize immunoglobulins, such as gamma-globulin. The important point here is that the immunity of insects lacks the molecular specifity of antibodies from mammals.

Induced Antibacterial Proteins

Wounding or injection of certain bacteria some insects induces the biosynthesis of antibacterial proteins called cecropins. These proteins were first isolated from the hemolymph of *Hyalophora cecropia*, from whence the term cecropin was coined. Molecules with similar structure have since been isolated from the waxmoth *Galleria*, the tobacco horn worm *Manduca sexta*, several species of Diptera and from a beetle. Not all insects appear to produce cecropins, because a cricket, a cockroach and a locuts do not produce them. The blood-sucking bug *Rhodnius* produces antibacterial proteins that differ from cecropins. *H. cecropia* produces cecropins A, B and D as well as several minor cecropins, C, E, and F, which differ from the major cecropin by only a single amino acid substitution.

All of these proteins are about 4000 in molecular weight, each with 35 to 37 amino acids. There is a very high level of homology among these proteins, all of which feature long hydrophobic reaches in the C-terminal portions of the molecule. The very close homology among these proteins is consistent with the idea that they have arisen through gene duplication.

The cecropins are not active against many, but not all, bacteria. Cecropins A and B are active against a large number of gram-positive and gram-negative species, including some that are occasional insect pathogens and some that are obligate insect pathogens and others that are non-pathogenic species. None of the cecropins are active against *B. theringeinsis*.

Antibacterial Actions

Cecropins lyse bacterial cell membranes; they also inhibit proline uptake and cause leaky membranes. These actions at the level of procaryotic membranes are consistent with the structures of cecropins. They are all cylindrical, amphipathic molecules with long hydrophobic regions on one end. In effect, these protein act like detergents. Synthesis of proteins that are analoges to cecropins show that activity against a broad spectrum of bacteria species is lost when amino acid substitutions that disrupt the alpha-helix structure are made.

The genes that code for cecropins have been cloned. These molecular biological approaches give insight into the biosynthesis of the cecropins. The gene for cecropin B, for example, codes for 26 amino acids in the N-terminus of the protein that are removed during post-transcriptional processing to produce the mature cecropin B. These are removed in 3 steps, first 22 residues are removed in one cleavage, which produces a pro-cecropin. The pro-cecropin is further processed by two sequential cleavages that yield the final product.

Attacins

A second set of antibacterial proteins are called attacins. These also have been isolated from *H. cecropia* and three other moth species. Attacins are considerably larger than cecropins, made up with over 180 amino acids. The antibacterial activity of these proteins covers a rather narrow spectrum compared to the cecropins, with high activity only against a few species of bacteria in the alimentary canal. Their site of action is at the outer bacterial membrane. These proteins appear to work in synergism with the cecropins and with lysozyme so that the three types of immune proteins can work together.

Lysozyme

Peptidoglycan is a large, bag-shaped macromolecule that encases bacterial membranes. It is universally present in bacteria. Minor variations of peptidoglycan occur in among different bacterial species. Lysozyme is an enzyme that hydrolyzes the glycosidic bonds of peptidoglycan, and thereby causes bacterial cell lysis. This enzyme constitutively occurs in low levels in hemolymph of most insects. Lysozyme is rapidly induced immediately upon bacterial infection.

Still Other Antibacterial Proteins

Antibacterial proteins widely occur in vertebrate and invertebrate animals. It is not too surprizing, then, to learn that representatives of various insect orders produce different antibactieral proteins. The attacins and cecropins are known mostly from lepidopterans, although they have been detected in other insects and even in intestine of pigs.

Honeybees produce another family of such proteins that are known as apidaecins. These are small peptides, molecular weight about 2100. Apidaecins do not appear to disturb bacterial membranes, and it is not yet clear how they do work. It is speculated that these are bacteriostatic, rather than bacteriocidal, proteins. Royalisin is another antibacterial peptide is found in the royal jelly of honeybees. This is also an amphipathic protein, and its mode of action may be similar to cecropins. It is a small molecule, molecular weight estimated at 5523.

Dipterans also produce induced antibacterial proteins. Two peptides from *Phormia* are called defensin and diptericin. Another dipteran peptide is a male-specific antibacterial product from *Drosophila melanogaster*. This peptide is known as andropin, and it thought to protect seminal fluid and the male reproductive tract against microbial infections. One of the most exciting findings in insect immunology is the discovery of an insect immune protein called hemolin. This protein belongs to the immunoglobulin superfamily, that is, it shares sequence homologies to mammalian immunoglobulins. It is thought that this protein is one of the first proteins to appear in the hemolymph of the giant silkmoth *H. cecropia*. This protein binds to the surface of bacteria, and it is likely that the binding may be the first step in the insect immune response.

Bacterial Resistance to Insect Immunity

With respect to evolutionary ecology, our discussion of insect immunity offers a sort of physiological perspective on host-parasite relationships. Within the context of these relationships, we can imagine strong evolutionary pressure on the microbes to develop mechanisms that circumvent immunity reactions with hosts. Two such mechanisms are known in some of the bacteria that invade insects. These are called passive resistence and active resistence. Please note that resistence in this context refers to the parasite side of host-parasite relationships. Passive resistence is related to the structures of some bacterial envelopes. These envelopes happen to be resistent to antibacterial activity of immune proteins. In these cases the resistence is a feature of outer membranes, rather than something bacteria might actively produce. There appear to be mutants of insect-pathogenic bacteria that have lost this sort of passive protection and are comcominantly less virulent.

Active resistence refers to active biochemical breakdown of immunity, rather than a sort of side-stepping as seen in passive resistence. B. theringeinsis, for example, produces an immune inhibitor called InA that works by proteolytic breakdown of cecropins and attacins. A beautifully complex relationship occurs in a nematode with a symbiotic bacterium appropriately named *Xenorhabdus nematophilus*. This bacterium is sensitive to two cecropins, A and B. However, the nematode produces an immune inhibitor enzyme that digests cecropins and attacins, thereby rendering them without activity.

One co-evolutionary response to these bacterial proteases that inactivate the insect anti-bacterial proteins may be another protein produced by some insects that inhibit the bacterial proteases. A molecule known as alpha2-macroglobulin has been identified in the hemolymph of some arthropods, but not yet from insects. Alpha2-macroglobulins are protease inhibitors, well known from mammals. These inhibitors do not work by competitively inhibiting the proteases, which is the most general mechanism of inhibiting proteases.

Instead, these molecules form "cages" around protease enzymes. The cages inhibit protease activity by screening larger molecules out. Smaller peptides are not screened out, so they can get into the active sites of the proteases, where they can be hydrolyzed. This forms a sort of selective inhibition of proteases. It is speculated that the alpha2-macroglobulins are a regular feature of arthropod hemolymph. They are thought to clear protease activity from hemolymph by collecting the proteases, then delivering them to hemocytes where they participate in a endocytotic protease clearance pathway that is similar to protease clearance pathways known in mammals.

The Biological Significance of Immunity

There are over two hundred species of bacteria that live in one or another association with insects. These can be categorized with repect to their pathogenicity:

- harmless forms that are not pathogenic
- occasional pathogens
- obligate pathogens.

Induced insect immune systems do not have antibacterial activity against all, or even, most pathogenic species. *B. theringensis* is a good example of a pathogenic species that is not troubled by immune systems. On the other hand, as far as can be known, insect immune sytems are effective against all non-pathogenic systems. What sense can we make of this? One suggestion is that insect immunity has more to do with regulating population dynamics of the non-pathogenic bacteria that naturally occur in insect bodies. My speculation is that many pathogenic bacterial species have secondarily lost protective and resistence mechanisms such that they have become labile to the induced proteins.

Other Humoral Immunity Factors

We most often regard phenyloxidases in terms of sclerotization, or tanning, of cuticle. These enzymes are found within cuticle and in hemolymph of insects. In hemolymph phenyloxidases are involved in melanin synthesis by way of quinones. Phenyloxidases are very active enzymes, and the intermediates and products of phenyloxidase reactions can be harmful to animals. These enzymes are nearly always found in a pro-phenyloxidase form. Phenyloxidases are activated in response to bacterial infections, and melanin, a tanned insoluble material is deposited around clumps of bacteria. A similar reaction occurs at the site of wounding in insect integument.

Activation of phenyloxidase has been studied in a few insects. This usually involves a signal from the surface of bacterial walls, such as the putative beta-1,3-glucan receptor. Conformational changes in this receptor activate a serine protease called prophenyloxidase activating enzyme. The activating enzyme removes a 5-kD peptide from prophenyloxidase, which yields phenyloxidase. These sort of cascade reactions are common biochemical activating mechanisms. Another important reaction cascade is the blood-clotting reaction in mammals.

Lectins are agglutinating molecules that are well known from plants and many classes of animals. Lectins have been purified from the hemolymph of many insect species, including a meat fly, a cricket and a grasshopper. A possible function of some lectins may be agglutination of microbes, but the this has been clearly shown in only a single case.

Insect Immunology: Hemocytic Immunity

As we mentioned last meeting, insects defend themselves from challenges by microorganisms with a combination of humoral and cellular responses. The distinction between these two response categories is rather artificial for a couple of reasons. First, hemocytes can release some of the protein factors that follow from challenges. Hemocytes produce what we regard as humoral responses. Second, although infections elicit a complex of multifaceted responses, the responses are coordinated and integrated into a single theme: host defence. Nevertheless, it is very convenient to treat hemocytic and humoral immunity as separate topics. One point of separation is that hemocytic responses are rapid. They occur almost immediately an insect is invaded by microorganisms. Humoral responses are much slower, they are usually not seen until some hours after infection. These longer-term responses may be more important in sustaining the hemocytic host defence reactions initiated earlier in the infection cycle.

Circulating cells are regularly found in the hemolymph of insects. These cells are known as hemocytes. Hemocytes are involved in conferring a cellular, or hemocytic, immunity to insects. These cells are formed in haemopoietic organs that are variously well developed in different insect groups. In the cricket *Gryllus* haemopoietic organs are well developed, distinct organs found in the second and third abdominal segments along either side of the heart. In other insects haemopoietic organs are amorphous clumps of cells irregularly distributed along the dorsal vessel.

You might enjoy a quote on the classification of insect hemocytes: "Classification of insect hemocytes is an issue that stimulates an instantaneous loss of good manners and an induction of bad temper at any conference on insect physiology or immunity". Given detectable levels of disagreement on the subject, a list of some hemocytes and their function follows: Prohemocytes (PRs) are thought to be the stem cells from which all hemocytes arise.

Plasmatocytes (PLs) are larger cells involved in major immunological actions. PLs are capable of phagocytosis, a kind of entocytosis of smaller foreigh cells. Phagocytosis usually follows low doses of invading yeast and bacteria. PLs are also involved in encapsulation of clumps of invading microbes. There are typically more PLs than other types of hemocytes.

Granulocytes (GRs) are characterized by the presences of many intracellular granules. Electron microscope studies have revealed differences among the granules in these cells. There are structureless and structured granules. The biochemistry of the different types of granules is not yet understood. GRs are major players in encapsulation reactions.

Other hemocytes include spherulocytes, oenocytoids and adipohemocytes. These cells are regarded as "nonimmunocompetent" by some authors, although there is considerable ground to disagree with this idea. Gupta prefers to use the term "immunocytes" for GRs and PLs and "hemocytes" for all the other types. He uses these terms to stress the idea that GRs and PLs make up most of the immune responses. Spherulocytes may be involved in silk formation, melanization and secretion of some hemolymph proteins. They may also be important in hemolymph clotting responses to wounds. Oenocytoids may contribute to the final stages of cuticle formation. Other authors suggested that these cells may recognize "non-self".

Defence Mechanisms

Again, hemocytic defence responses occur immediately insects are invaded by microorganisms. GRs and PLs are responsible for a number of cellular immune reactions. They are able to phagocytize small biologic particles such as bacteria and yeast. Phagocytization refers to the process of cellular internalization of foreign invaders. Phagocytosis is not the same as killing the microbial cells. Microbes are secondarily killed after phagoccytosis.

GRs and oenocytoids are attracted to sites of wounds, where they play major roles in coagulating hemolymph to stop bleedings. We know virtually nothing about the signalling process that is responsible for moving cells to wound sites.

Another defence mechanism is nodule formation. Nodules are aggregates of hemocytes that entrap invading microbes. This is a particularly effective means of clearing the hemolymph, and it is considered to be of greater importance than phagocytosis in clearing large doses of bacteria. There appears to be a sequence of trapping by GRs, followed by aggregation by PLs. The precise sequence of events will remain a point of controversy. In addition to bacteria, viral polyhedrons, fungal spores and protozoans are taken up in nodules.

When organisms or clumps of organisms are too big to be phagocytized or to form nodules, they may be encapsulated. Encapsulation is the sequesterization of these organisms or clumps with multilayered aggregates of hemocytes. This also involves release of coagulum, in a response similar to wound-repair. The coagulum probably aides in forming layers of cells. Encapsulation was first observed by the Russian zoologist Elie Metchnikoff in 1892. He placed rose thorns into a starfish larva. The next day he saw that the thorns were completely surrounded by tiny cells. Metchnikoff called these cells phagocytes from the Greek words for eating and cells. Metchnikoff eventually moved from the University of Odessa to the Pasteur Institute, and later shared the Nobel Prize for his studies of cellular defences in mammals and invertebrates.

We are not sure of the killing mechanisms that hemocytes elaborate. Some of the suggested mechanisms include peroxidation, melanization, and lysozymes. By comparison to our scant knowledge of invertebrate immunity, very much is understood about the killing mechanisms of mammalian host-defence cells. There are two sorts of mechanisms. One sort does not depend on oxygen. These include protease, esterase, acid-phosphatase and lysoszyme mechanisms.

Another sort does depend on oxygen, including formation of singlet oxygen and superoxide anions, which are very reactive and very toxic species. Work on killing mechanisms of insect hemocytes will produce important and very novel insights into how insects protect themselves.

We recently addressed the question of how the early, hemocytic responses to infection are mediated. What we wanted to know was what biochemical signals occurred between recognition of a bacterial infection and the hemocytic responses that follow. Since many aspects of host defence in mammals are mediated prostaglandins and related eicosanoids, we formed the hypothesis that these molecules are also crucial mediators of immunity in insects. The central experimental design was to inhibit various of the eicosanoid biosynthetic pathways, then observe the effects of inhibition on the insects' ability to remove bacterial infections from their hemolymph.

Since hemocytic responses occur in the first few minutes post-infection (PI), we could dissect the humoral responses from hemocytic responses by doing the experiments within the first hour PI. In the first experiments, cells of the insect pathogenic bacterium *Serratia marcescens* were injected into 5th instar larvae of the tobacco hornworm *Manduca sexta*.

Test animals were first injected with the phospholipase A2 inhibitor dexamethasone, and control animals were first treated with pure ethanol. Hemolymph samples were withdrawn at 15, 30 45 and 60 minutes PI. The samples were diluted slightly, then streaked onto agar plates. The plates were incubated for about 40 hours, then the number of red colonies on each plate were counted. Until these data are scanned into the Web site, you'll have to look at the original publication, or use the handouts.

The line connected by square boxes represents hemolymph samples from control insects: there were no red colonies on these plates, indicating that the insects had cleared all or most of the bacteria from their hemolymph. The line connected by X's represents hemolymph samples from test insects. As early as 15 minutes PI, there was a statistically significant increase in recovered bacteria. The number of recovered colonies increased over the first hour PI.

We tested the effects of several doses of dexamethasone by injecting 4 doses in different groups of insects. The results indicate that more bacterial colonies were recovered with increasing doses of dexamethasone.

The bacteria were used in these experiments are insect pathogens. We reasoned that the dexamethasone decreased the ability of the larvae to clear bacteria from their hemolymph. Since these bacteria are pathogens, the decreased clearance from hemolymph should translate into increased mortality from these infections. We tested this idea by injecting the same 4 doses that were just described into 4 groups of larvae. After 14 hours we counted the number of surviving larvae. As expected, increasing dosages of dexamethasone were associated with decreasing larval survival at 14 hours.

Dexamethasone inhibits the activity of phospholipase A2. This in turn inhibits eicosanoid biosynthesis because substrate is not available to the eicosanoid-biosynthetic pathways. We conducted experiments to test the possibility that arachidonic acid would rescue the larvae from the effects of dexamethasone. These kinds of experiments are called "end product reversal experiments", or something like that. That when larvae were treated with dexamethasone, and with arachidonic acid, larval survivorship was a great as survival in the control group.

That phospholipase A2 activity is necessary for all eicosanoid biosynthesis. We wondered if one eicosanoid biosynthetic pathway might be more important than another in mediating insect immune responses.

To investigate this point we treated groups of larvae with specific inhibitors of phospholipase A2 (dexamethasone), cyclooxygenase (indomethacin), lipoxygenase (esculetin), and indomethacin and esculetin together. We also observed the effects of maleic acid, an antioxidant compound that does not interfere with the major eicosanoid biosynthesis pathways.

After treating the larvae with inhibitors, we infected them with bacteria. After 2.5 hours, we withdrew hemolymph samples, and plated them on agar plates. We then counted the number of bacterial colonies on each plate. This is a rigorous experiment because if any bacterial colonies were recovered, even a single colony, the individual was counted as positive.

These data show that at least 1 colony was recovered from about 20% of the control larvae. Colonies were recovered from about 80% of the phospholipase A2-inhibited larvae. Bacterial colonies were also recovered from the cyclooxygenase-inhibited and the lipoxygenase-inhibited larvae, but from about 60%, rather than the 80% of phospholipase A2-inhibited larvae. These findings indicated that both pathways were involved in the immune response to infections. This idea was supported by the results from the group that was treated with both inhibitors at the same time: they were not different from the dexamethasone effects.

We concluded that products of cycloxygenase and lipoxygenase pathways were involved in mediating cellular responses to bacterial infections in this insect.

The last comment, of course, opens questions of identifying specific cellular actions that are mediated by eicosanoids. As mentioned earlier, nodule formation is responsible for clearing the bulk of bacterial infections from circulation. We developed the hypothesis that eicosanoids mediate the nodulation reaction to bacterial infections.

Using a line of research similar to the one just described, we found that treating tobacco hornworms with eicosanoid biosynthesis inhibitors just prior to infecting them with bacteria resulted in severe impairment of nodulation. The influence of the inhibitors is expressed in a dose-dependent manner and can be reversed by eicosanoid-precursor polyunsaturated fatty acids. Again, we concluded that products of cycloxygenase and lipoxygenase pathways were involved in mediating nodulation responses to bacterial infections in this insect. Finally, we began testing the broader hypothesis that eicosanoids mediate nodulation in other insect species. So far, we have tested these species:

Order	*Common name*
Lepidoptera	Tobacco hornworms
	True armyworms
	Black cutworms
	Silkworms
Coleoptera	King beetles-a large tenebrionid
Orthoptera	Crickets
	Cockroaches

This work serves to support the broader hypothesis that eicosanoids mediate nodulation responses in all insects that defend themselves from bacterial infections by forming nodules. Since publishing our original findings, other laboratory groups have taken up this line of work. We now know that eicosanoids mediate phagocytosis and nodulation in greater waxmoths.

Endocrinology of Insect Development

The events that take place in the egg stage of insect life cycles are known as embryonic development. Upon emerging from eggs, juvenile insects embark on an excursion of post-embryonic development that will eventually take them to their adult forms. Post-embryonic development is characterized by dramatic increases in size and by formation of new anatomical structures characteristic of adults. Argued from an ecological point of view, juvenile stages of insects are devoted to eating, growing, and accumulating energy reserves while adult stages are devoted to migrations, mate-finding and reproduction. This appears to be particularly true among some insects, such as certain species of sphinx moths, which do not feed during adulthood. But as generalizations, these broadly sweeping remarks are not strictly true. Development of reproductive organs begins early in larval life in many insect species, so even if an insect is not actually reproducing, it is preparing for reproduction quite early in life. Many adult insects spend quite a lot of time in feeding behaviours. The value of broad generalizations is that we can appreciate the major thrust of insect life stages. Certainly larval insects are largely given over to eating and growing.

Insects are invested in a more or less rigid exoskeleton, and a generally accepted wisdom is the physical constraints of a least certain of the very rigid parts of exoskeletons inhibit growth beyond set limits. Here is an example of the overall increase in weight during several weeks of development. The potential to increase in size during post-embryonic development, in this species from a few milligrams at hatching to nearly 2 grams. Fully hardened cuticle does not expand, so growth of the hardest parts can only take place during molts. The softer parts of insects can expand. The softer cuticle often occurs in folds that can be expanded, and sometimes the cuticle itself can be stretched. The soft parts of an insect grow rather continuously, while the hard parts grow in steps.

Growth can refer to an increase in cell number or an increase in cell size. The development and expression of cuticle depends on the cells of the epidermis. Epidermal cell numbers increase prior to molting in many insect species. In larvae of many Diptera, growth is due exclusively to increased size of the epidermal cells. Similarly, growth of internal organs can proceed by increased cell size or increased cell number. Both forms of growth occur in the mosquito *Aedes*. Cell numbers in the nervous system and fat body increase, but most other tissues have a constant number of cells. Growth occurs by cell enlargement in these tissues.

Malpighian tubules are separated into primary Malpighian tubules and secondary Malpighian tubules. The primary tubules arise from cells of the proctodeum during embryonic development. The secondary tubules emerge during post-embryonic development. The secondary tubules are reduced to buds during each molt, then develop by cell division, following by cell enlargement.

Let us turn to the endocrinology of molting. We will consider five hormones in the regulation of postembryonic development, plus a newly discovered, sixth hormone. The actions of the five hormones are summarize.

Prothoracicotropic hormone (PTTH) is synthesized in neurosecretory cells of the brain. It is stored and released from the corpora cardiaca. Again, PTTH is released from the neurohemal portion of the CC. PTTH stimulates the prothoracic glands to release ecdysone into the hemolymph. The prothoracic glands have other names. You will see the prothoracic glands called by other names, including ventral glands and ecdysial glands. In higher Diptera it is part of the ring gland. Ecdysone is not the active molting hormone. Various tissues, including the fat body convert ecdysone to 20-hydroxyecdysone, the active form of molting hormone, shown here.

The cells of the epidermis respond to 20-hydroxyecdysone with initiation of the process of molting. Apolysis refers to the detachment of old cuticle from the epidermis, the first step in molting. Molting fluid is actively transported into the space between the old cuticle and the epidermis. Epidermal growth by cell division may occur during this time. The next step involves depositing layer of new cuticle. This is followed by activation of enzymes in the molting fluid that function to digest inner layers of the old cuticle.

The amount of juvenile hormone (JH) released from the corpora allata (CA) determines the form of the new cuticle that is deposited. When JH is present in high concentrations, the new cuticle is larval; when JH is present in low concentrations, the new cuticle is pupal. Adult cuticle is formed in the absence of

JH. Sir Vincent Wigglesworth, who was knighted for his research in medically important insects, first suggested the idea of an "inhibitory hormone" in the 1930's. Working with penultimate instars of the blood-sucking bug, *Rodnius*, Wigglesworth showed that if bugs were decapitated 5 days after a blood meal, they molted into the last larval instar. If the bugs were decapitated 3 days after a blood meal, they molted into precocious, but miniature adults. The Harvard biologist Carrol Williams gave JH its name. Williams worked with *Cecropia* moths. The adult moths were parabiotically joined to diapausing pupae. Parabiotically is a relatively rare word for people who work on mammalian systems. Parabiotic animals are joined in such a way that they share hemolymph circulation. Wigglesworth was famous for his "parabiotic trains", in which many *Rhodnius* nymphs were joined in long trains; molting would progress rearward down the trains. Williams' idea was to try to prolong the life span of the adult moths, which do not feed, by providing nutrients from the pupae. The interesting thing was this: when the pupae were joined to adult females, the pupae would continue normal development and emerge as adults. Alternatively, when the pupae happened to be joined to male adults, they molted into another PUPAL stage. A series of now famous experiments finally showed that males of some moths, including *Cecropia* moths, accumulate a blood-borne factor in their abdomens that exerts a "status quo" effect on pupal moths. Williams named the factor juvenile hormone because it caused molting into a second pupal stage.

Ecdysis refers to shedding the old cuticle, and is the basis of the name ecdysone for molting hormone. We note that apolysis and ecdysis are separate processes. In some insects shedding of the pupal skin by adults is triggered by another brain hormone called eclosion hormone (EH). EH is synthesized in the median neurosecretory cells of the brain, and stored in the neurohemal portion of the CC. It is released from the CC under regulation of a circadian rhythm. EH stimulates the abdominal nervous system to play out a "hard-wired" behavioural program that helps the adults emerge from the pupal case.

We consider a figure, showing the influence of EH on nerve firing in an in vitro preparation of the tobacco hornworm ventral nerve cord. EH was added to the preparation 22 minutes before the first spike occurred. The early spikes (A and B) represent the rotational bursts that begin ecdysis. The later spikes (C and D) represent peristaltic bursts. Jim Truman and Lynn Riddiford, then at Harvard, are responsible for the discovery of EH, and its role in helping insects emerge from the pupal case. Truman and Riddiford married, and took up dual positions in the Department of Zoology at the University of Washington in Seattle.

Let us now introduce a newly discovered hormone in the physiology of insect development. While molting appears to be rather abrupt, the process is long and detailed. As molting progresses, the inner layers of the old cuticle are digested away. Then a soft new cuticle is deposited. When these developmental events are complete, the shedding processes begin, and the new cuticle is expanded and hardened. The key point is noting the shedding processes are behavioural, and the role of EH is to release a pre-programmed sequence of behaviours. We now know the release of the ecdysial behaviours is more complicated.

Last year Zitnan and colleagues, then at UC Riverside, discovered the existence of a second ecdysial hormone, called Mas-ETH. Mas comes from *Manduca sexta*, and ETH stands for ecdysis-triggering hormone. ETH is a 26-amino acid peptide. It is produced by a system of epitracheal glands. In *M. sexta* there are 8 pairs of epitracheal glands, each located on the outer surface of the large tracheal tube immediately adjacent to each spiracle. The hormone is produced and released from Inka cells. Inka is the name of a fairy goddess, known from the Tatra Mountains. Zitnan takes great inspiration from Inka, and he named the cells in her honour.

Ecdysis is now thought to follow from the sequential release of two hormones. EH is released from the brain, and it primes the Inka cells to release ETH. The Inka cells release ETH, and the ETH acts on the abdominal ganglia to release the behaviours required for ecdysis. But why do we need two hormones to release ecdysial behaviours? Ecdysis involves many behaviours that are unique to ecdysis. Successful ecdysis requires that all behaviours are coordinated. Failures in coordination result in deformities and death: each ecdysis episode is a crucial event in an insect's life. The Inka cells are positioned near each spiracle, where tracheal linings are withdrawn at ecdysis. These linings are fragile, and if torn the tracheal supply to that region is blocked. It is thought the Inka cells are in best position to serve as a last inspection site before the insect is irrevocably committed to going through ecdysis.

The physiology of molting involves apolysis, depositing new cuticle and ecdysis. The circulating titre of JH at the time of the molt determines the developmental fate of the insect. Here is an example of the relationship between ecdysone and JH through 4th and 5th larval instars, the pre-pupal wandering stage and pupal stage of the tobacco hornworm. The solid line shows ecdysone titres. Ecdysone titre spikes a day or so before molting into the 5th instar, then again just before the wandering stage as a prepupa. There is a small increase in ecdysone titre before the pupal molt. Ecdysone titres reach a higher level and remain elevated during the pupal stage. The dotted line represents JH titres. There are high titres of JH in the 4th instar, which are lower in the 5th instar. A small spike of JH is associated with the wandering phase, and no JH is associated with the pupal stage.

JH is also active in adult stages of insect life. JH plays an important role of regulation of vitellogenin synthesis in adult female fat bodies. JH acts directly upon follicular epithelia of ovaries to facilitate uptake of yolk proteins. JH is also active in adult males, where it regulates development of reproductive tract accessory glands.

Hormones are transported in circulating hemolymph. Since the hemolymph bathes virtually every cell in the insect body, circulating hormones have the potential to come into contact with all tissues. Again, specific receptor sites form the connection between a circulating hormone and particular target cell.

JH and ecdysteroids are not water soluble, and they are associated with specific high-affinity transport proteins. These transport proteins, then, are an important link in the overall physiology of endocrine regulation of biochemical and physiological events. These proteins are also partly responsible for regulating circulating hormone levels. They can act by protecting the hormone from enzyme degradation and by preventing rapid loss of hormone from the hemolymph through excretion.

The titre or levels of circulating hormones are regulated at several points. Biosynthesis, release, transport and degradation of hormones are regulated events. Biosynthesis of ecdysone depends upon PTTH from the brain. JH biosynthesis is partly regulated by nervous signals from the brain to the CA. Ecdysone is degraded to a biologically inactive form, 3-dehydro-alpha-ecdysone, and 20-hydroxyecdysone may similarly be oxidized to a 3-dehydro form.

There are two major pathways that change JH into biologically inactive forms. The minor pathway is hydration of the epoxide functionality to the diol. The most important mechanism is hydrolysis of the JH ester to its respective acid. JH can be broken down by general carboxyesterases, enzymes that occur throughout the insect body and by JH-specific esterases. The JH-carrier protein protects circulating JH from the general esterases, but not from the JH-specific esterase.

The endocrinology of insect reproduction is not without controversy. Mosquitoes elaborate an egg development neurohormone (EDNH) that is produced in the brain. EDNH is released after females take a blood meal. The hormone acts on the pre-vitellogenic ovary, and stimulates release of ecdysone, which is converted to 20-hydroxyecdysone.

The 20-hydroxyecdysone then acts on the fat body, which results in vitellogenin synthesis, and development of the vitellogenic ovary. The interpretation of these data is that in mosquitoes the ecdysteroid produced by the mosquito ovary acts on the fat body to stimulate synthesis of yolk proteins. This, of course, runs counter to the conventional wisdom that JH is released from the CA, then acts on the fat body to stimulate yolk protein synthesis. Among people who are interested in the topic, there are two camps: one group absolutely disagrees with this idea.

Other reproductive hormone actions have been postulated for some insect species. Ken Davey and his co-workers in Canada suggested that there is an internal feedback control of ovarian development in the bug Rodnius. In this insect an antigonadotropin is thought to inhibit the action of JH on ovarian follicles to prevent from developing patency. This is regarded as a form of direct ovarian inhibition.

Indirect ovarian inhibition is thought to occur in several species of Diptera. The general scheme is that the follicle cells produce an oostatic hormone that acts on the neuroendocrine system, causing a reduction of JH.

Anatomy and Physiology of Reproductive Systems

Two features of insects that facilitate their great ecological impact are their ability to move over impressive distances and their amazing potential for reproduction. We often speak of reproduction in a single, simple word. The apparent simplicity is misleading. Reproduction entails a great number of successive processe, each of which occur in the correct order one after another. There is not enough time to discuss all aspects of reproduction, but we might list these steps as follows: [1] sex determination, [2] gonial mitoses, [3] meiosis, [4] differentiation of reproductive organs (i.e., gonads, ducts and accessory glands), [5] spermatogenesis, [6] ovarian development, including previtellogenesis and vetellogenesis, [7] accessory gland functioning, [8] pheromone production, [9] sexual behaviour, [10] mating, [11] ovulation, [12] oviposition, [13] incubation of embryos or larvae, and [14] parturition. Another important feature of insect reproduction is these many steps are often tightly tied to various environmental cues, such as photoperiod and temperature.

Even though we can reduce a complicated process to a list of intricate steps, it is important to stress that living, real insects orchestrate these steps in harmony with cues from their environment. From this point of view, reproduction necessarily involves sensory physiology, afferant nervous transmission of sensory information, and central nervous system integration of information from environments. These neurobiological aspects are to be conceptually justaposed with anatomy and physiology of organs and glands that directly subserve reproduction.

Sex Determination

Sexual characteristics of mammals are influenced by a combination of genetical and hormonal events. We may think mammalian sex determination is a strictly genetical event. The situation is otherwise. There is in humans, for example, a condition called testicular feminization in which individuals with the chromosomal make up of males look like females during juvenile and adult phases of life. This condition is brought on by a failure to produce androgenic hormones or hormone receptors. We take the point that sex determination in mammals is not strictly genetics. Combined genetic and hormone action was similarly thought to control sexual differentiation of gonads, accessory glands and other structures relating to mating and reproduction in insects. Reproduction is a fundamental aspect of insect biology, and a great deal of research effort has been spent on the question of sex determination. Many experiments consisted of transplanting gonads from male to female and female to male in waxmoths, cross transplanting wing discs in a *Gryllus* cricket, and several other species, and cross transplanting genital discs in *Tenebrio*, among several other species. Results from all of these experiments show that hormones do not regulate sex determination in insects.

Of course, there is an counter example. Again, one of the most interesting features of insect physiology is the variety in how things are done. Sex determination in adults of the glowworm *Lampyris noctiluca* appears to be under hormonal regulation. We will consider this case because it is, so far, the only known example of hormonal influence of sex determination in insects. Sexual differentiation of male and female gonads begins during the 4th larval instar. This differences are subtle at this stage, but in males cells in the apical tissue begin to divide, whereas in females it is cells in basal tissue of the gonads that begin to divide. When testes were transplante into females earlier than 5th instar larvae, they induced a transformation, or a masculinization, in the female recipients. Testes could not produce such effects in female-determinted larvae after late in the 5th instar. Converse experiments did not produce feminization in young males. Hence, ovaries of this species can not feminize males, but testes can transform ovaries of pre-5th instar females into testes.

When ovaries were transplanted into males, they became masculinized; this was so, albeit to a lesser degree, even when ovaries were placed into males that had been testectomized. These results suggest that another tissue is also involved in ovarial transformation. Other experiments with testectomized males that were further treated by ablation of various endocrine organs finally showed that two hormones are involved in male differentiation: an androgenic hormone from the apical tissue of testes and a neurohormone from pars intercerebralis, which triggers apical tissue differentiation.

The contemporary wisdom holds that for most insects sex differentiation is not regulated by hormones, with the single known exception just described.

Anatomy of Male Reproductive Systems

There is usually a pair of testes in males, as shown for *Tenebrio*. Sometimes the testes appear fused as shown for *Locusta*. Vasa deferentia (pleural of vas deferens) connect testes to seminal vesicles, which are in turn connected to an ejaculatory duct. In *Locusta* there are 15 pairs of glands, not counting the seminal vesicles. Some of the accessory glands are given individual names, such as white gland, or opalescent gland. In other species, such as a cricket I am familiar with, there are many accessory glands, and they all look alike, so individual glands can be readily identified.

The ejaculatory duct is ectodermal in origin, and therefore cuticle-lined. Some accessory glands are also derived from ectodermal tissue, and these are called ectadenia (pleural of ectadene); ectadenia are found in Coleoptera and some other groups. Other accessory glands derive from mesodermal tissue and are called mesadenia; mesadenia are known in Orthoptera and other groups. *Tenebrio* is an example of an insect with both sorts of accessory glands. Testes are made up of follicles or testis tubes. Numbers of follicles varies among insect groups from as few as one to over 100 per testis. Testes of some insects, such as Tenebrio shown earlier are a series of lobes, each of which contains follicles. Some beetles have over a dozen lobes in each testes. Often follicles are connected to the vas deferens by a short tube called vas eferens, as shown. In other cases, particularly in Lepidoptera, follicles are not completely separated and they share a common opening to vas deferens. Usually vasa deferentia dialate, or widen, to form the seminal vesicle just before joining the ejaculatory duct.

Testicular Physiology

In the far end of follicles is the germarium. Here germ cells divide to produce spermatogonia. Spermatogonia become encysted in groups of various sizes. As more spermatogonia are produced, they force older ones further down the follicle. Hence, there is a gradient of maturation, with mature sperm in the proximal part of follicles near vasa deferentia, and newly formed spermatogonia in the distal end of follicles. Between these points, people think of threee zone of development along follicles. In Zone I, spermatogonia divide and transform into larger spermatocytes. Zone II is regarded as an area of maturation. Here, each spermatocyte two meiotic divisions to produce spermatids. Finally, in Zone III, spermatids develop into spermatozoa.

The overall path looks something like this:

Germ cells ———> spermatogonia —————>
spermatocytes ———> spermatids ————> sperm

You may note very little is known about regulation of spermiogenesis. More recently, though, it has been firmly established that ecdysone is necessary for spermiogenesis in many species.

Anatomy of Female Reproductive Systems

The major features of female reproductive systems, in this case from a locust and from a dipteran. In the general scheme, there are usually two ovaries, from which lateral oviducts join a common genital chamber. In insect species that transfer sperm by direct intromission of a penis, rather than by a spermatophore, the genital chamber is called a bursa copulatrix or vagina. A sperm storage organ, or spermatheca, is connected to the genital chamber by a spermathecal duct. As in males, female reproductive tracts often have accessory glands. Whereas testes are composed of follicles, ovaries are made up of ovarioles. Ovaries have from a few to over 100 ovarioles, depending upon species and, in locusts, on population densities. As usual, there are extremes, and some termites have over 2,000 ovarioles per ovary while certain aphids have only a single ovariole in one functioning ovary.

Three types of ovaries are recognized, based on the arrangement of nurse cells. These are panoistic, telotrophic and polytrophic. *Panoistic ovarioles* do not have specialized nurse cells. For attachment with abdomens, oogonium, and layers of external sheath, including tunica propria. Here we see oogonia can develop into young oocytes, and then into primary oocytes. As in testis follicles, there is a spectrum of increasing maturation toward the proximal end of ovarioles. Panoistic ovarioles are regarded as the most primitive type, and they are usually found in more primitive orders such as Odonata and Plecoptera.

T*elotrophic* ovarioles have trophic tissue in addition to oogonia an oocytes in the germarium. Trophic cells and oogonia derive from germ cells in the germarium. As oocytes mature, they move toward the proximal end of ovarioles. The maturing oocytes maintain contact with the core of tropic tissue by way of a nutritive cord that grows longer as oocytes move toward the proximal end of ovarioles. The nutritive cords break during vitogenesis, and follicle cells form a continuous layer around the oocyte.

Polytrophic ovarioles have trophocytes wrapped up with each oocyte in a follicle. These ovarioles are found in most holometabolous insects. In the germarium, oogonia divide into an oocyte and a trophocyte, but the two cell types maintain a cytoplamic link. Further cell divisions can occur so that a given ooctye may be associated with several trophocytes.

Telotrophic and polytrophic ovarioles, those characterized by having nurse cells, are collectively called meriostic ovarioles. We use the term trophic frequently in this discussion, in telotrophic and polytrophic ovarioles and in trophic tissue and trophocytes. Its enough to make you think these cells feed something to the oocytes, but that is not quite what happens. The principle role of trophic tissue is to supply large amounts of RNA, both ribosomal RNA and transfer RNA, to oocytes. This is thought to support rapid oocyte growth.

Ovariole Physiology

An ovariole usually holds a series of oocytes, each further along in development as they near the proximal end. The germarium is at the distal end, and here stem line oogonia derive from germ cells. The oogonia divide into two daughter cells, one of which becomes an oocyte and the other retains it stem cell function. As an oocyte leaves a germarium is becomes invested in a prefollicular tissue that later forms the follicle cell layer surrounding each oocyte. The number of follicle cells increases as oocytes grow. In Drosophila, for example, they increase from about 80 to over 1200. During yolk deposition the follicle cells do not divide, but insteat form a flattened epithelium. The proximal portion of an ovariole is called the vitellarium, which in mature insects is by far the greater part of an ovariole. Yolk deposition, or vitellogenesis occurs in the proximal part. Oocytes vastly increase in size during vitellogenesis. Yolk is categorized as protein yolk and lipid yolk. Juvenile hormone stimulates fat bodies to synthesize and export vitellogens, or yolk proteins. JH also causes gaps in follicular cells, a condition called patency by Ken Davy. This allows free exchange between oocytes and hemolymph, and the oocytes rapidly take up proteins from circulation. Lipid yolk is also derived from fat body. It is thought that early in development only phospholipids are taken up, then triacylglycerol and phospholipids. The lipids probably provide metabolic energy and maternally derived essential fatty acids to developing embryos.

When vitellogenesis is completed the outer layer of the oocytes usually develops. This is called the vitelline membrane which is partly secreted by follicle cells and partly by oocytes. We already know that ovulation refers to the movement of mature oocytes from the ovarioles to the common oviduct.

Biochemical Roles of Vitamins in Energy Metabolism

This lecture is meant to replace the planned lecture on regulating feeding behaviours. We will use this period to extend our treatment of nutrition by adding details on the biochemical roles of some vitamins. This will also serve as a review of energy metabolism, meant to help prepare for the final examination.

We recall that muscle contraction involves interactions between thick myosin filaments and thin actin filaments. These interactions occur by way of cross bridges formed by heads of the myosin molecules. Mg++-dependent Myosin ATPase is responsible for the conformational changes in the orientation of the myosin cross bridges that result in the filaments sliding past on another. The sliding filaments result in a shortening of the Z-band to Z-band distance.

Tracheoles functionally, but not topologically, penetrate deep into rapidly metabolizing muscle fibres such that oxygen can be delivered nearly directly to muscle mitochondria, sometimes called sarcosomes. The key function of the gas exchange is to support energy metabolism during flight, or any other energy-intensive activity. By now you should have an idea of where, morphologically and biochemically, oxygen is consumed and where carbon dioxide is produced. We can revisit energy metabolism by looking back at the chart that shows types and forms of fuels.

We can also recall the metabolic pathways that yield energy to support cell work. These include glycolysis, Kreb's Cycle, the Electron Transport Chain and ß-oxidation. A physiologist might say, barring some needling details, your grasp on energy metabolism is virtually unrivalled in the western hemisphere.

The sugars, lipids and protein components that insects and other organisms consume in the course of their feeding activities are called macronutrients, not because they are particularly large or complex molecules, but because they are required in relatively large quantities. In addition to these macronutrients, insects require a host of micronutrients, so called because they are needed in more or less trace levels in the body. Some of these micronutrients are VITAMINS, which can be defined by three points. First, they are found in trace to low quantities in the body. Second, they are involved in specific metabolic functions. Third, they are not synthesized in the body. If they are synthesized, the synthetic rates do not support the organismal needs.

Some of the vitamins are directly involved in energy metabolism, acting within the pathways within our line of familiarity. We are going to look at these vitamins, and at the specific metabolic functions they perform. This is one of the areas of biochemistry that integrates common features of many different organisms. These metabolic functions are just as vital to mammals, including human beings. Generally speaking, we recognize two major groups of vitamins-the fat soluble and the water soluble vitamins. The fat soluble group includes vitamins A, D, E, and K. The water soluble group includes the B-complex and vitamin C.

We proceed with vitamins we have already seen in action. The structure of riboflavin is shown first. The key feature here is that the vitamin-that is, the trace substance that can not be manufactured within the body-is merely a component of a larger molecule that plays a specific role as a coenzyme. Before we discuss the role of FAD, let us introduce nicotinamide, also shown in its vitamin form and in its physiological form of nicotinamide adenine dinucleotide, similarly abbreviated as NAD.

Both FAD and NAD function as electron carriers. When an electron pair is transferred to these molecules, either one or two hydrogens are also transferred. FAD has accommodation for two hydrogens while NAD can accept one hydrogen. In NAD, then, an electron pair and one hydrogen are transferred, with a second hydrogen released into the medium. This subtle chemical difference is why the reduced form of FAD is written FADH2 while the reduced form of NAD is written NADH + H+. These compounds are particularly appropriate electron carriers because they readily accept an electron pair and they also readily donate an electron pair. We saw FAD act in the Kreb's cycle. FAD accepts an electron pair from succinate, becoming FADH2. FADH2 carries the electron pair to the electron transport chain, where it releases its electrons at such position in the chain that two ATPs are subsequently formed. That NAD operates as an electron carrier at three places in the Kreb's cycle, carrying electrons from isocitrate, alpha-ketogluterate and from malate to the electron transport chain. Here the electrons are loaded onto the chain such that three ATPs are formed.

These examples illustrate a general feature of co-enzymes: they all provide some chemistry to enzymes that the proteins can not perform on their own. As coenzymes, FAD and NAD provide the ability to accept and donate electrons.

Riboflavin and nicotinamide are necessary components of these co-enzymes, and these components can not be made in the body of insects. For perspective, recall from your biochemistry that NAD acts as a co-enzyme in more than 250 enzyme functions of this sort: carrying electrons from one place to another.

Our next water soluble vitamin is B-1 or thiamin, shown in its vitamin form and in its physiologically active form as thiamin pyrophosphate, or what is sometimes called cocarboxylase. This vitamin is quite often referred to as TPP. As a point of history, it was because of the amine group in this compound that accessory food factors were first called vitamines, an abbreviation of vital amines. Specifically, TPP is a coenzyme in the removal of carbon dioxide from various compounds. An important reaction is this one:

Pyruvate--> Acetyl CoA + CO_2

Recall that this reaction is carried out by pyruvate decarboxylase. This reaction removes the carbon dioxide moiety from pyruvate, and it connects the glycolytic pathway to the Kreb's cycle.

Returning to the Kreb's cycle, note that carbon dioxide is released at two points in the cycle. The active form of thiamin, TPP, also functions in the release of this second carbon dioxide. We collect the point that some decarboxylations, or carbon dioxide removal reactions, do not depend the activity of this co-enzyme.

We have now seen three vitamins-two function as electron carriers, essentially connecting the Kreb's cycle to the electron transport chain and the third functions in decarboxylation reactions. In one decarboxylation reaction, specifically pyruvate decarboxylase, TPP also helps connect the glycolytic pathway to the Kreb's cycle. The transport of 2-carbon acetyl units, done by Co-enzyme A or CoA, is still another key link between glycolysis and the Kreb's cycle. Recall that oxidation of fatty acids in the ß-oxidation pathway is also connected to the Kreb's cycle by transfer of acetyl CoA. This serves to underscore the importance of vitamin B-3, or pantothenic acid. This is shown in its vitamin form and as a component of Co-enzyme A. These four vitamins-nicotinamide, riboflavin, thiamin and pantothenic acid-are directly involved in energy metabolism; they act as electron carriers, in decarboxylations and as an essential component of CoA.

Insects require other water soluble vitamins in their diets, including biotin and folic acid. These do not function specifically in energy metabolism. Nonetheless, we want to briefly consider their roles in cellular metabolism. The chemical structure of biotin was worded out in 1942, and first synthesized in 1943. Biotin had been independently discovered by at least two groups, and it went under a variety of names, including vitamin H and protective factor X. Biotin functions as a carbon dioxide carrier in carbon dioxide fixation reactions. Here is an example:

Acetyl CoA + CO_2 —— > Malonyl CoA ——> Fatty Acids

Folic acid is actually a complex of several biologically active compounds. The formal term is pteroylglutamic acid, and it is composed of pteroic acid plus one to seven gluatmate residues. Folic acid is required for transferring single carbon units. It is essential in metabolism of some amino acids, and is probably most well known for its critical role in synthesis of thymine.

Without folic acid, DNA synthesis is slowed or stopped. In the absence of DNA synthesis, production of some cells, such as red blood cells, is also slowed or stopped. This is particularly important for cells that undergo rapid turnover, such as red blood cells or insect hemocytes.

With respect to the water soluble vitamins, insects appear to be closely aligned with known vertebrate nutritional requirements: both groups require about the same vitamins for the same biochemical reactions. This is reasonable because the major energy yielding pathways are similar in both groups.

The fat soluble vitamins, vitamins A, D, E, and K, are not involved in the oxidative pathways we have discussed. Vitamins D and K are involved in functions occurring in vertebrates, but not invertebrates. Vitamin D is involved in bone metabolism, and vitamin K is involved in the vertebrate blood clotting cascade. Vitamin K was discovered in 1929 by a Danish scientist, Henrik Dam, who later won a Nobel prize for the work. The K comes from the Danish word koagulation. The long-standing wisdom holds that insects or other invertebrates do not require vitamins D and K.

Vitamin A is involved in many mammalian tissues, mostly in ways not completely understood. One well-appreciated action of this nutrient is its role in vision. We will discuss the physiology of vision in a later lecture, in which the role of vitamin A will be addressed in detail. For now we will just make the point that vitamin A is required for insect vision. Vitamin E is also thought to be dietarily essential in at least some insects. Vitamin E is essential for fertility in rats, and apparently in male crickets. The specific roles of vitamin E are not understood, although its most well known activity is its action as an antioxidant. Vitamin E is thought to interact with certain free radicals, and thereby prevent oxidative damage to cells. Whether vitamin E is required by all insects is difficult to say. I speculate it is required, however, the requirement is very hard to detect because the vitamin is stored in tissues and turns over only very slowly.

Now there is one last problem in energy metabolism. Referring to the glycolytic pathway again, please note that at one point NAD picks up a pair of electrons to become NADH + H+. There are two points to be made: One, if the electrons were not subsequently donated from the NADH, cells would eventually become depleted of NAD-and energy metabolism would come to a screeching halt right at this step. So it is important to recycle this electron carrier to keep energy metabolism rolling along.

The second point is that this pair of electrons represents a certain amount of potential energy in the form of additional ATPs if the electron pair could be shuffled over to the electron transport chain. Stated in problematic terms, we have this problem: the outer mitochondrial membrane is not permeable to NAD. There are two pools of NAD in most cells. One is the inner mitochondrial pool and the other is the cytoplasmic pool. Since the electron transport chain is tucked away in the inner membrane of the mitochondria, how is the electron pair, associated with the cytoplasmic pool of NAD, to be delivered to the electron transport chain, or is it delivered at all?

Let us consider the operation of the glycerol-3-phosphate shuttle. On the left is the glycolytic pathway. This is in the cytosol fraction of the cell, where the NAD accepts an electron pair and becomes NADH (plus H^+). The electron pair is transferred to dihydroxyacetone phosphate (DHAP).

This step regenerates the NAD, thereby allowing the glycolytic pathway to continue its important operations. Reducing the DHAP by addition of electrons converts it from DHAP to glycerol-3-phosphate. This newly formed compound easily migrates across the outer mitochondrial membrane. Again, the membrane is a barrier to NAD flux. Once in the mitochondria, the electron pair is donated to the electron transport chain at the level of FAD, which converts the glycerol-3-phosphate back into DHAP. DHAP is free to wander back out into the cytosol. Then two points are covered by the glycerol-3-phosphate shuttle: the NAD is regenerated and the potential ATPs are realized.

This shuttle is best known from insect flight muscles. There are other pathways to account for moving reducing equivalents from cytosol into mitochondria, one of which is called the MALATE SHUTTLE because malate is formed by oxidation of oxaloacetic acid.

The malate moves into the mitochondria, where it is converted to fumarte and then into succinate. This is a more complex shuttle, and I once speculated at a scientific meeting that the glycerol-3-phosphate shuttle is much faster than the malate shuttle, hence more appropriate to very rapidly metabolizing tissues, such as insect flight muscle.

Sensory Physiology: Insect Vision

Vision is the perception of light. Insects perceive light through three classes of sensory organs. Most adult insects have a single pair of compound eyes. Larvae of hemimetabolous insects and most adults have usually three simple eyes called ocelli. These are typically located on the dorsum of the head capsule, and they are sometimes called dorsal ocelli. Larvae of holometabolous insects do not have compound eyes. These insects perceive forms, to a limited extent through stemmata located on either side of the head.

Compound Eyes

The compound eyes are prominent features of adult insects. The eyes occupy a fairly large portion of the surface of insect heads, and they facilitate a rather wide field of vision. Compounds eyes are not present in all insects. They are reduced or absent in parasitic forms, many soil insects, and in some species that live in very dark places, such as caves. The basic unit of compound eyes is the ommatidium. Ommatidia vary in size and number among insect groups. At one extreme, dragonflies have thousands of ommatidia in each compound eye; many insects have far fewer, and at the other extreme, the workers of the ant species *Pomera punctatissima* have only one ommatidium in each eye. The sizes of ommatidia vary from about 5 to 40 microns in diameter. Sizes vary among species and even within a single compound eye. In some dragonflies, for example, the dorsal units are considerably larger than the ventral ones.

The ommatidium rests on a basement membrane. The height of the eye comes from the long retinula cells and secondary pigment cells. The corneagen cells (sometimes called primary pigment cells) rests atop the long retinula and secondary pigment cells. The crystalline cone lies within the corneagen cells. The surface of the ommatidium is covered with the corneal lens. The corneal lens is a specialized part of insect cuticle. It is secreted by the corneagen cells, which are specialized epidermal cells. The crystalline cone is a clear, intracellular structure synthesized within Sempter cells. The primary pigment cells surround the Sempter cells. Not all compounds eyes feature a crystalline cone; those that do are called eucone eyes.

The corneal lens and crystalline cone make up the optical part of ommatidia. The sensory elements are just under the optical apparatus. The retinula cells are long neurons. Part of each retinula cell is a specialized area known as a rhabdomere. The rhabdom indicated in the figure is a collection of up to eight rhabdomeres. A nerve axon from each retinula cell projects through the basement membrane into the optic nerve. Ommatidia are functionally isolated because the retinula cells are surrounded by the secondary pigment cells.

Compound eyes are commonly categorized as either apposition eyes or superposition eyes. There is no separation between the corneal layer and the photoreceptors in apposition eyes. There is a clear space between the two units in superpositions eyes. Some authors use the expression "clear zone eyes" for the superposition eyes. Superposition eyes are generally found in crepuscular and nocturnal insects, while apposition eyes occur in diurnal insects.

The crystalline cone couples the lens and photoreceptors in apposition eyes. Some higher Diptera have pseudocone eyes. Crystalline cones do not occur in pseudocone eyes, and optical coupling is by means of a gelatinous substance that is contained in a two-celled structure. Still another arrangement is found in some apposition eyes that lack solid cones or gelatinous pseudocones. In these acone eyes four flat transparent cells are found in place of the cones. In superposition eyes the space between the corneal lines and the photoreceptors is traversed by crystalline tracts. These are thought to act as wave-guides that direct light to the photoreceptors. The eyes of some insects feature a tracheal tapetum. Tracheal tapeta are known from several Lepidoptera. Small tracheae run parallel with the longer cells of the ommatidia. It has been suggested that these structures function as interference reflector filters.

Light sensitivity of insect eyes varies according to the state of light-adaptation or dark-adaptation of the eyes. In bright light, the eye is less sensitive to light, and the eye is regarded as light-adapted. Maximum sensitivity occurs in darker conditions, when the eye is fully dark-adapted. Some insect eyes can change light sensitivity by about three orders of magnitude within a few minutes of changing light conditions. There are a number of mechanisms of adaptation to light and dark conditions. One mechanism relates to the biochemistry of vision. The visual pigment is broken down by interaction with light. In daylight the breakdown rate can equal or exceed the replacement rate. This leads to decreased light sensitivity, and partly explains light adaptation. In the dark, the visual pigment accumulates, and the insect becomes dark adapted.

A second mechanism of adaptation is the movement of screening pigments. Screening pigments are located in the retinula cells. In light-adapted eyes granules of screening pigments surround the rhabdom. These pigments absorb light, and they have the effect of optically isolating individual ommatidia. The endoplasmic reticulum of dark-adapted eyes forms large, clear vesicles. This forms a clear space around the rhabdom, so that a greater amount of light falls upon the rhabdom.

Vision involves the transduction of light energy into a bioelectric signal within the nervous system. The first events in this process take place in the retinula cells. The fine structure of rhabdomeres consists of thousands of closely packed tubules. These tubules are about 500 A in diameter and one micron long. They are aligned at right angles to the long axis of the rhabdom.

The visual pigments occur mainly in these rhabdomeric microvilli. It has been suggested that the small diameter of each microvillus inhibits free rotation of visual pigments. This specific orientation may be the molecular basis of insects' sensitivity to polarized light. Photobiological processes occur in a narrow band of the electromagnetic spectrum between 300 and 700 nm. Photons in this region of the spectrum have enough energy for photochemical interactions, but not enough energy to disrupt macromolecules. Visual pigments initiate vision by absorbing light in this region. These pigments are a class of membrane bound proteins known as opsins that are conjugated with a chromophore. Visual pigments whose chromophore is retinal are called rhodopsins. The visual pigments of all invertebrates, including insects, crustaceans and squids, are all rhodopsins. There are a number of geometric isomers of retinal.

The 11-cis isomer is the chromophore of native rhodopsin. Free retinal absorbs light strongly at about 380 nm. When 11-cis-retinal is linked to opsin to form rhodopsin, the main absorption peak is shifted. The maximum absorption values of different rhodopsins range from about 345 nm (ultraviolet) to as high as 610 nm (red). The absorbing properties of any given rhodopsin is related to the disposition of charged opsin groups around the chromophore. These interactions between the side-groups of the amino acids within the opsins and the chromophores are thought to modify the p electron orbital of the chromophores, and thereby modify the absorbance spectra of rhodopsins.

When the chromophore of rhodopsin absorbs a photon, the chromophore undergoes an isomerization from the 11-cis to the all-trans configuration. The rhodopsins of insects and other invertebrates differ from vertebrate rhodopsins on the point of the chemical events that follow the isomerization step. Vertebrate rhodopsins go through a series of spectrally distinct intermediates.

The process is called bleaching because the absorption shifts from visible to ultraviolet wavelengths, which we can not see. The vertebrate process can lead to hydrolysis of the opsin and chromophore. Invertebrate rhodopsins generally do not bleach. Light transforms the rhodopsins to a stable intermediate. The isomerized 11-cis chromophore stays in place in the microvilli. The stable intermediates have spectral properties similar to metarhodopsin I of vertebrates, which has earned them the term metarhodopsins.

The mechanism of transducing light-induced chemical changes in rhodopsins into electrical signals remains the major unresolved issue in understanding the physiology of light receptors. Several milliseconds are required before the absorption of light can be recorded as electrical activity in the receptor cell membrane. Certainly, the phototransduction event occurs during the transition from rhodopsin to metarhodopsin. We can not say much more about the detailed mechanisms of transducing light into bioelectrical signals.

A key requirement in the chemistry of vision is that metarhodopsin, or the bleached pigment of vertebrates, must be reconverted to rhodopsins. In the eyes of vertebrates, this regeneration is the product of a complex biochemical pathway. The pathway involves isomerization back to the 11-cis configuration and assembly with the opsins. Again, invertebrates differ from vertebrates on this aspect of photochemistry. Light is absorbed by rhodopsin and metarhodopsin in invertebrates. Absorption of light transforms metarhodopsin back to rhodopsin. The pigments exist in two stable conformations, and light causes them to switch back and fro from one state to the other. This regeneration is essential because conformational changes in rhodopsin, but not in metarhopsin, leads to phototransduction.

Experiments on insect behaviour indicate that many insect species can distinguish colours. Colour vision depends on the ability to discriminate between light of different wavelengths. Such discrimination is possible because compound eyes have photopigments with differing spectral properties. Many insects have two rhodopsins, one with maximum absorption in ultraviolet wavelengths and one with maximum absorption in green wavelengths. Some species have a third pigment that absorbs maximally in the blue region. A few Lepidoptera have four visual pigments.

Dorsal Ocelli

Dorsal ocelli occur in larvae of hemimetabolous insects and in nearly all adults. Here is the structure of a dorsal ocellus. Although there are variations in structure, a typical ocellus has a single lens that is usually rather thickened. Most ocelli feature a large number, often hundreds, of retinula cells. Rhabdomeres of several cells combine to form a number of rhabdoms. Axons project from the retinula cells through the basement membrane and they end in a synaptic plexus behind the eye. Light causes a sustained depolarization of the retinula cells, however, the biological roles of ocelli remain unknown. It is generally agreed that ocelli allow only poor, if any, perception of form. In some orthopterans the ocelli are active in orientation to a light source.

Stemmata

The larvae of holometabolous insects have stemmata on the sides of the head capsule. Many larvae have a single stemma on either side, but the number can be as many as six on a side. Here is a sketch of a section through a stemma. Each stemma has a cuticular lens and a crystalline cone distal to the sense cells. Stemmata generally do not produce clear images, but most caterpillars can discriminate some shapes and they can orientate themselves with respect to boundaries.

Insect Olfaction

Host Seeking Behaviour in Female Mosquitoes

Many female mosquitoes actively seek potential hosts in the semi-darkness of the evening or the darkness of night, and most mosquitoes depend upon olfactory, rather than visual, information to locate potential hosts. This is true, also, for diurnal females. While visual cues are quite important for long-range detection of potential hosts, virtually all female mosquitoes depend upon olfactory cues for locating hosts in short-ranges. This is just one example of the significance of olfactory information in the biology of insects, and we might consider this in a little more detail.

Carbon dioxide (CO_2) is a key component of host odors. CO_2 stimulates two aspects of host seeking behaviour. CO_2 activates mosquitoes, that is, it induces take off and helps maintain flight behaviour. For a second role, CO_2 also helps female mosquitoes orient themselves with respect to a potential host. While most olfactory receptors are located on the antennae of insects, the mosquito CO_2 receptors are located on the labial palps. The olfactory responses to CO_2 are integrated with other olfactory information to produce the host seeking behaviours of female mosquitoes. Recognizing that information from the palps is integrated with information from the antennae, we note the integration of information from the peripheral system is integrated in the central nervous system. We also note CO_2 is a necessary component of host odors. In the absence of CO_2, other odor components can not stimulate host orientation behaviours.

Lactic acid (LA) is probably the primary host odor to which female mosquitoes orient themselves during flight. Lactic acid is a ubiquitous product of energy metabolism. While high levels of LA are produced under anaerobic conditions, low levels are also produced under aerobic conditions. LA evaporates from the skin of humans and other mammals.

Mosquitoes have LA receptors on the sensilla basiconica, or grooved pegs, on their antennae. The LA receptors are specific to lactic acid, with very little activity in response to other small, organic acids. In humans, the LA flux rate is about 1.5 x 10^{-12} moles/second, which is well within the range of sensitivity of the mosquito LA receptor. There are actually two LA receptors in the antennae of female mosquitoes. One of the receptors is stimulatory and the other is inhibitory, that is, there is increased frequency of nerve impulses in one and a reduced frequency in the other. The biological significance of the inhibitory lactic acid receptor is not well understood.

The physiology of the LA receptor is of some interest because the expression of the LA receptor sensitivity coincides with host-seeking behaviour. Newly emerged mosquitoes, and recently blood-fed mosquitoes do not seek potential hosts for blood meals. In both situations, the lack of host seeking behaviour is related to activity of the LA receptors. Newly emerged females do not respond to LA, however, as the females develop, there is a progressive increase in LA sensitivity. By the time females are of host-seeking age, usually 24 or more hours after emergence, the LA receptors are sensitive to very low concentrations of LA. It was once though the sensitivity of the LA receptors was tied to juvenile hormone levels, because LA sensitivity is correlated to pre-vitellogenic ovarian development. But the correlation is not physiologically linked: females whose corpora allata were removed at adult emergence undergo a normal progression of LA sensitivity.

Female mosquitoes are often interrupted in the course of taking a blood meal. Because they may not receive a full meal, females require a threshold of blood before they can be said to have a blood meal. Small meals below the threshold do not stimulate ovarian development, nor influence subsequent feeding behaviours. Once they acquire a threshold level of blood, usually 2.5 ml, females substantially modify their host seeking behaviours. For the first few hours after a blood meal, the anterior portions of female abdomens distended, and the distention is recorded by stretch receptors. The firing of these receptors inhibits host seeking behaviour. This inhibition is not mediated by changes in peripheral nervous system, and in particular, the sensitivity of LA receptors does not change. By the time the distention is relieved, however, an endocrinological mechanism continues to inhibit host seeking behaviour. An ovarian factor, thought to be ecdysteriod, primes the fat body to release another host seeking inhibitory factor. The fat body factor, thought to be a small peptide, sustains the inhibition of host seeking behaviour through its influence on the LA receptor: the sensitivity of the LA receptor is shifted to a very insensitive state. Hence, in the second post-feeding phase, host seeking behaviour is inhibited by shifting the sensitivity of a peripheral receptor. Information about LA and CO is integrated with other information, as well. Quite sensitive thermoreceptors are present in the sensilla coeloconica, and efforts to generate an artificial host odor reveal that temperature is one component of host odor used to orient with respect to potential hosts.

Olfactory Reception of Sex Pheromones in Moths

We considered two aspects of sex pheromones in insects. First, we saw that sex pheromone biosynthesis can be understood as slight variations on fatty acid metabolism in many insect species. Second, we considered the physiological regulation of sex pheromone biosynthesis, seeing again that insects have evolved a variety of mechanisms to achieve a single end, in this case a timely release of sex pheromones. Sensing and responding to sex pheromones lies on other side of biosynthesizing and releasing airborne sex pheromones. Let us look at the olfactory reception of sex pheromones in male moths.

The males of many moth species are activated to sustained upwind flight by airborne sex pheromones released by females of the same species. Males can detect sex pheromones released from females over an impressive distance, from a few meters to a few thousand meters. The physiological mechanism of sex pheromone reception is not much different from the mechanisms of detecting host odors, although we have quite a bit more information from male moths.

The physiology of detecting a sex pheromone probably begins with releasing the hormone into the air. Pheromone molecules are deposited from the air onto the surface of antennal hairs. The molecules sort of flow through pores in the hairs and make their way into a gel that surrounds the dendrites of antennal nerves. The gel is mostly protein, specifically proteins comprising the pheromone binding proteins of male moths. Glen Prestwich and some of his colleagues have tried to understand the molecular biology of these pheromone binding proteins. Finally, pheromone molecule, coupled to its pheromone binding protein, interacts with cell surface receptor sites located on the outer surface of the pheromone receptor neuron. Again, a neuron is just a nerve cell, and in antennae these nerve cells are arranged with the dendrite projecting within the shaft of the hair, cell bodies at the base of the antennae, and axons projecting from the cell body to the brain.

The receptors for pheromones are conceptually similar to all receptor sites. They are transmembrane proteins with amazing specificity for particular pheromone molecules. Pheromone receptors interact in a physiologically relevant way with one pheromone, and only one pheromone. To be just a little more precise, the receptors interact with one and only one component of a pheromone. The pheromone-receptor interaction generates a depolarization of the neuron membrane, and if the depolarization is strong enough an action potential flows from the cell body at the base of the antenna to the brain. By strong enough depolarization, we are referring to multiple interactions between several pheromone molecules and their receptors.

The neurophysiology and behavioural physiology of pheromone reception is a very important field of inquiry, and many people are attracted to the work. Leaders in the field include John Hildebrand who works on males of the sphinx moth, *Manduca sexta*. John moved from Columbia University to the University of Arizona in the 1980's, where he successfully lured Reg Chapman from UC Berkeley. Tom Baker was a student of Wendell Roelofs at Cornell and is now head of the Department of Entomology at Iowa State, is also a leader in understanding the integration of pheromone signals. Tom moved from UC Riverside to Iowa. Thanks to the work of Glen Prestwich, we have information on the interaction of pheromone components with their specific receptors. Many people have published work on the electrophysiological patterns of action potentials that follow from interactions with pheromones. The contemporary questions revolve around how blends of pheromones are detected and integrated.

The main idea comes from a beautiful combination of morphological and electrophysiological research. Male, but not female, moths have a structure at the base of their antennae known as the macroglomerular complex (MGC). The MGC receives information exclusively from sex pheromone receptor neurons. Thousands of sex pheromone receptor neurons are present on the antennae of male moths. Action potentials from these neurons are directed, by way of the antennal nerve, to the MGC. The breakthrough in understanding the detection and integration of pheromone blends came with appreciating the structure of the MGC. The MGC is subdivided into about 7 subcompartments, each of which receives olfactory information from exactly one component of a pheromone blend.

Let us consider a two-component pheromone system, taken from a recent article by Tom Baker and one of his post-docs. While pheromone molecules can and will be deposited on all antennal sinsillae, molecules of each each component will bind to specific binding proteins and interact with specific receptors tuned to a specific molecule. The pheromone-receptor interactions generate depolarizations, which can lead to action potentials. The action potentials run down to the appropriate subcompartment of the MGC. Two possibilities can follow from here. In one scenario, the axons form synaptic connections with interneurons which serve as integrating interneurons. These can integrate signals about relative intensities of each of the two components, and project an integrated signal deeper into the central nervous system, eventually to the protocerebrum. Alternatively, interneurons specific to each receptor neuron can project deeper into the central nervous system, where integration of the information can take place.

This basic scheme works for multi-component pheromones. The frontier of this research field lies in understanding how multi-component pheromones are integrated into a single piece of information that will direct the behaviours of male moths.

Again, this is a contemporary and very active area of insect science. We can complete our discussion of this area by asking how to determine a direct link between pheromone receptor neurons and specific subcompartments of the MGC. The key lies in modifying some neuron staining procedures from the early days of neural anatomy.

One procedure is called "back-filling". A friend of mine at UC Berkeley used this to work out the neural anatomy of selected nerves in a cricket. The goal is to place a nerve fibre into a solution of cobalt dye, then let the dye move through the axons, and follow the path. The work is done in 3-dimensional preparations, and it is much more difficult than it sounds. The technique has been modified to that a sensillum is cut, then placed in an electrode filled with a cobalt solution. The sensillum was stimulated with a pheromone component, and the cobalt solution moved down the sensillum and into the MGC. In many of these preparations, we see specific staining in separate subcompartments of the MGC.

What is Forensic Entomology?

Forensic entomology is the use of insect knowledge in the investigation of crimes or even civil disputes. It is one of the many tools of forensic science. You will see that forensic entomology is not only a useful tool to decide how long human remains have been undetected, but forensic entomology can also be used to find out whether the corpse has been moved after death, the cause of death, and also solve cases of contraband trafficking.

Forensic entomology can be said to be the application of the study of insects and other arthropods to legal issues. It can be divided in three subfields: urban, stored-product and medicolegal.

Medicolegal forensic entomology includes arthropod involvement in events such as murder, suicide and rape, but also includes physical abuse and contraband trafficking. Since the earth is a predominantly arthropod world, it is not uncommon that we mere humans comes in contact with these creatures. They make the world go round, as they pollinate, eat other arthropods, eat living plants and trees, dead plants and trees, living vertebrates, dead vertebrates and vertebrate dung and urine and a lot of other things.

The feature with arthropods that are most important in medicolegal forensic entomology is that they are important carrion feeders, that is they eat dead vertebrate bodies, including man. Thus they perform a valuable recycling of organic matter in our ecosystem.

One of the first groups of insects that arrive on a dead vertebrate is usually blowflies. Usually the female oviposits within two days after death of the vertebrate. Then the blowfly goes trough the following stadiums: egg, 1. instar larvae, 2. instar larvae, 3. instar larvae, prepupae, pupae within puparium, imago.

If we know how long it takes to reach the different stadiums in an insects life, we can calculate the time since the egg was laid. This calculation of the age of the insects can be considered as an estimate of the time of death. But even if the estimate of the insect age is correct, the death of the victim (usually) occurred before the eggs were laid. This period is quite variable and depends on temperature, time of day the death occurred, time in year the death occurred, whether the corpse is exposed or immersed in soil or water. As a general rule insects will lay eggs on a corpse within two days after the corpse is available for insects. Insects can also be of help in establishing whether the corpse has been moved after death, by comparing the local fauna around the body, and the fauna on the body.

In some instances, movement of suspects, goods, victims or suspect vehicles can be traced with the help of insects. Insects parts, or whole insects can for example be captured in different car parts, such as in radiators or tyre treads. By identifying the insects found, and plotting the distribution of each insect, as well as the biology of each species one can find the greatest degree of overlap, and describe the areas where the suspect has been.

Estimating Time of Death With Forensic Entomology

After the initial decay, and the body begins to smell, different types of insects are attracted to the dead body. The insects that usually arrives first is the Diptera, in particular the blow flies or Calliphoridae and the flesh flies or Sarcophagidae.

The females will lay their eggs on the body, especially around the natural orifices such as the nose, eyes(2), ears(2), anus, penis and vagina. If the body has wounds the eggs are also laid in such. Flesh flies do not lay eggs, but deposits larvae instead.

After some short time, depending on species, the egg hatches into a small larvae. This larvae lives on the dead tissue and grows fast. After a little time the larva molts, and reaches the second larval instar. Then it eats very much, and it molts to its third instar. When the larvae is fully grown it becomes restless and begins to wander. It is now in its prepupal stage. The prepupae then molts into a pupae, but keeps the third larval instars skin, which becomes the so-called puparium. Typically it takes between one week and two weeks from the egg to the pupae stage. The exact time depends on the species and the temperature in the surroundings. The theory behind estimating time of death, or rather the post mortem interval with the help of insects are very simple: since insects arrive on the body soon after death, estimating the age of the insects will also lead to an estimation of the time of death.

How to Estimate Age of Blowfly Eggs, Larvae, Pupae and Adults

Eggs: When blow flies oviposit, their eggs has come very short in their embryonic development. The eggs are approximate 2 mm in length. During the first eight hours or so there is little signs of development. This changes after that, and one can see the larvae through the chorion of the egg at the end of the egg stage. The egg stage typically lasts a day or so.

Larvae: The blowfly has three instars of larvae. The first instar is approximately 5 mm long after 1.8 days, the second instar is approximately 10 mm long after 2.5 days, the third instar is approximately 17 mm long after 4-5 days. Identifying the right instar is the easiest part, and is done relatively easy based on size of larvae, the size of the larva's mouth parts and morphology of the posterior spiracles. The time it takes to reach the different instars depends very much on microclimate, i.e. temperature and humidity.

Prepupae: At the end of the third instar the larva becomes restless and starts to move away from the body. The crop will gradually be emptied for blood, and the fat body will gradually obscure the internal features of the larvae. We say that the larva has become a prepupa. The prepupa is about 12 mm long, and is seen 8-12 days after oviposition.

Pupa: The prepupa gradually becomes a pupa, which darkens with age. The pupa which are about 9 mm in length are seen 18-24 days after oviposition. The presence of empty puparia should therefore tell the forensic entomologist that the person in question has been dead in more than approximately 20 days. Identification can be done based on the remaining mouth parts of the third instar larvae.

A more precise way to determine age of larvae and eggs is the use of rearing. For example: the body is found with masses of eggs on it, none have hatched. How long time is it since the eggs was oviposited? Note the time of the discovery, note the time when the first 1. instar larvae occur. Subtract the first occurrence time with the discovery time, call this time A. Rear the blow flies to adults, let them mate, let them lay eggs on raw beef liver under conditions similar to the crime scene, take the time from oviposition to the first occurrence of 1. instar larvae. Call this time B. By subtracting B-A, one gets C, which is an estimate of the time since oviposition to discovery. Similar calculations can be done for other instars as well. If one has good base-line data from before under different temperatures and for different species, one only needs to rear the flies to a stage where they can be identified, and that is the third stage or the adult stage.

One important biological phenomenon that occurs on cadavers are a succession of organisms that thrive on the different parts. E.g. beetles that specialize on bone, will have to wait until bone is exposed. Predatory rove beetles or parasites that feed on maggots will have to wait until the blow flies arrive and lay their eggs.

The succession on cadavers happens in a fairly predictable sequence and can be used in estimating time of death if the body has been lying around for some time.

The first group to arrive is blow flies, followed shortly by staphylinids. As putrefaction develops, more groups arrive at the scene, with most groups present just before the body is drying out due to seepage of liquids. After the body is drying out, dermestids, tineids and certain mites will be the dominant animal groups on the body, and blow flies will gradually vanish. Fauna changes in the soil can also be used to estimate time since death.

Several insects are specialized in living in very decayed dead bodies. One example is the cheese skipper, *Piophila casei*, where the larvae usually occurs 3-6 months after death. The cheese skipper is a well known pest of cheese and bacon worldwide, and has a cosmopolitan distribution. Adult cheese skippers may occur early after death, but larvae occurs later. The earliest observation on human remains are when the body is two months, and this was under excellent summer conditions. In 1898, Potter examined 150 graves, and found remains of *P. casei* in ten of them. These graves were from three to ten years old and three to six feet deep.

In temperate regions dead bodies often appear in spring, after the snow is gone. The forensic entomologist and the forensic pathologist must then try to determine whether the death occurred during the winter or before the snow set in. If the death occurred before November, it is possible to find dead insects in and on the body. By analysing the dead insect fauna, and estimating when the insects probably died (this can be found by looking at meteorological records). Another hint is when the different adults stop flying before the winter. For example: here in Norway, we have had some cases where the bodies have been found in the spring. In one case we found dead third stage blow fly larvae in the back of the mouth. The blow fly larva was of an species that is flying from May to October. It was from this concluded that the eggs probably was laid during October, and since it was relatively few larvae, probably late in October. In another case, we found several live insects on a dead body, and also many dead third stage larvae. The dead larvae was found on the stomach, the arms, the shoulders, and inside the head. We concluded that the live insects had colonized the dead body in the spring, and that the dead larvae had died during the winter. Based on the widespread occurrence of the larvae, we had to say it was likely that the body was colonized before October, probably in September.

If the death occurred in the winter things become difficult in outdoor settings, as very few insects are active in the winter. It is reported that larvae of the winter gnat, *Trichocera sp.* can develop on carrion in the winter. By estimating the age of these larvae, if present, it could be possible to estimate the PMI.

Finding the Cause of Death Using Forensic Entomology

In a crime investigation, there is not only of great interest to find out when a victim died, but also of interest to find out how the victim died, as this can be used to find the killer.

In some instances the insects themselves are the killers, in other instances the insects occurring on the carrion can shed a light on what happened when the victim died.

Wasps and bees, for example, can inject venom through a sting. Some people are sensitive and allergic to these venoms, and can die if not treated in time. One other important aspect of wasps and bees are their effect on drivers. Many car accidents are probably caused by some wasp, bee or bumble-bee coming through the window, causing hysteria, or a distraction from the road leading to a collision or other accidents. In some cases wasps and bees has been used as murder weapons, as in case where some parents had shut their infant in a room full of wasps, in order to get rid of it.

Poison can be traced in blood, urine, stomach contents, hair and nails. One other important source is maggots occurring on a corpse. After a while it will be impossible to sample stomach contents, urine and blood from the dead body, but it will still be possible to sample from maggots, empty puparia or larval skin cast.

Many of these chemicals will also influence the life-cycle of the maggot. For example will high dosages of cocain accelerate the development of some sarcophagids. Malathion, an insecticide, is commonly used in suicide, and is usually taken orally. Presence of malathion in the mouth may lead to a delay in the colonisation of the mouth. Presence of amitriptyline, an antidepressant, can prolong the developmental time with up to 77 hours, at least in one species of Sarcophagidae. Knowledge of drug use in the victim is therefore important not only in finding the death cause, but also in estimating the time of death.

The sites of blowfly infestation on the corpse may be important in determining the cause of death, or at least in reconstruction of events prior to death. For example:

if there have been trauma or mutilation of the body prior to death, this may lead to heavy infestation of other body parts than the usual sites when the victim is not mutilated. Under a knife attack, it is usual to guard oneself with arms in front of thorax and head. This may lead to injury on the lower part of the arm. After death, blowfly may oviposit in these wounds.

The usual sites of oviposition on dead humans are natural openings. Even here there is preference. Blowflies will most often lay their eggs in the facial region, and more seldom in the genitoanal region. If there is a sexual assault prior to death, leading to bleeding in the genitoanal region, blowflies will be more likely to oviposit in these regions. Therefore, if there is observed blowfly activity in the genitoanal region, one can start to suspect a sexual crime. This must of course be corroborated with other evidence as well. The interpretation of maggots in the anogenital region becomes very fuzzy after a few (4-5) days, as eggs will be oviposited in this region during the course of decay in the natural course of decomposition.

3

Decomposition of Forensic Entomological

One aspect of forensic entomology is the use of insect science in investigations of human deaths. A detailed background on the stages of decomposition and associated fauna is key to interpreting information such as postmortem interval. Forensic entomologists use the time of insect colonization in order to give the shortest estimate for the postmortem interval. This estimate is determined by using degree days to find when the first eggs were laid on the body. However, the time of death and time of insect colonization may not occur simultaneously. Colonization can be delayed by many factors, which is why exploring the external factors that affect decomposition and the colonization of fauna is vital when using entomological evidence in an investigation.

Decomposition

Decomposition is the process whereby bodily tissues are broken down into smaller molecules after death. The physical and chemical properties observed during decay are categorized into five stages: (1) fresh, (2) putrefaction, (3) black putrefaction, (4) butyric fermentation, and (5) dry decay. Seconds after death, insects are attracted to a body for food and to lay eggs in natural openings and orifices of the body. This process of decomposition best describes the relationship between insect colonization and the availability of different food sources throughout decay. The presence or absence of a particular insect, location on the body, and food source, can help determine forensically pertinent information about the body. Insects can provide significant insight to the manner of death, movement of the corpse from one site to another, and length of the postmortem interval. A decaying carcass provides "a temporarily, rapidly changing resource which supports a large, dynamic arthropod community." —M. Grassberger and C. Frank

Fresh

The fresh stage is the first phase of decomposition that begins approximately four minutes after death and lasts around three days until putrefaction. Autolysis initiates decay as digestive enzymes in the body begin to break down nearby cell membranes and digest the internal organs. The acidity in the body also increases due to lack of oxygen in tissues. The liver which has high enzyme content releases nutrient-rich fluids into the body. At this time, the body has little to no odor, the skin is quickly turning pale and showing signs of livor mortis. The body loses heat until it reaches ambient temperature, known as algor mortis. After several hours, fibres in the muscles of the body begin to bind together and stiffen. This is known as rigor mortis and is first seen in the smaller muscles of the body like the face. Insects will feed between the muscle during rigor mortis because of the lactic acid produced from the break down of the muscle.

The first necrophages observed on the body after death is the Calliphorid flies. The female green bottle fly is generally the first to colonize the body; the second is the Hairy Maggot Blowfly, *Chrysomya rufifacies.* Other fly families generally present during this stage are Sarcophagidae, Piophilidae, and Muscidae.

Predators of both immature and adult flies are prevalent at the beginning stages of decomposition. These are important factors to consider when determining the insect colonization time. *Chrysomya rufifacies*, the second fly to colonize, facultatively predates on fly larvae in its second and third larval instar. *Saprinus pennsylvanicus,* is a predaceous beetle in the United States that feeds on the early fly larvae. Parasitic wasps, fire ants, and other insects also predate on fly larvae. The parasitic wasp uses the flies' pupae as a nest for their eggs; upon hatching, the wasp larvae feed on the host maggots or pupae. Once the wasp larvae have killed the host, they use the fly remains to pupate into mature wasps.

Putrefaction

Following initial decay, approximately 4 to 10 days after death, the body begins the second stage of decay called putrefaction. During putrefaction, bacteria and other microorganisms continue to anaerobically metabolize the soft tissues of the body. This breakdown of tissue releases gases into the body and causes an increased internal pressure, which results in a bloated corpse. These gases commonly consist of hydrogen sulfide, carbon dioxide, methane, cadaverine, ammonia, sulfur dioxide, hydrogen, and putrescine. According to Arpad Vass, a greenish discoloration of skin due to the formation of sulfhaemoglobin in settled blood often signals the onset of the putrefaction stage. As a result of the gas accumulation and the progression of decay, the corpse begins to emit a smell. This smell is a putrid odor that tends to attract an increasing number and various species of flies, beetles, and other arthropods.

During putrefaction, maggot habitation begins as flies continue to arrive and oviposit in the orifices and natural openings of the corpse. Moreover, the first blowflies that arrived on the corpse during initial decay, have produced larva that are in their first and second instars. Clown beetles from the family Histeridae such as *Hister quadrinotatus* and *Hister seqkovi* in the United States are also attracted during the bloat stage and can be found underneath the decaying body.

Black putrefaction

Black putrefaction, also known as active decay, happens about 10 to 25 days after death. A good indicator of black putrefaction is a strong odor and black coloration of the corpse. The bloating begins to subside as the skin starts to peel back and break from the large amount of gas and fluids produced. This makes the body appear flat. The breaking of the skin allows insects and other consumers better access to the inside of the body. This open access to the inside of the body cavity aids in increasing the rate at which viscera and other soft tissue decay. Any remaining flesh still on the body has a soft creamy consistency. The nails on the fingers and toes of the body will readily disconnect.

Saponification is a processes that can slow down the rate of decay and lengthen the putrefaction process for years. This process has also been called adipocere or grave wax. In humid, warm conditions, saponification may be seen on the body in the form of an oily, waxy yellowish–white substance. This process occurs in the fats of animals and humans when ester bonds, such as those in triglycerides, are exposed to a chemical base. In this case, water from moisture in the environment acts as the chemical base and breaks the ester bonds of triglycerides of fat deposits. When these bonds break, fatty acids are released, which can form new bonds. What these free fatty acids bond to can influence the rate of decay and the characteristics of the waxy substance formed. This is aided by putrefactive bacteria, such as Clostridium perfringens, to help make fat turn into a soapy substance.

At this stage, the maggot mass starts to decrease as most of the maggots have reached their third molt and will start to leave the body in order to pupate in the soil. These maggots may live along side the larvae and adults of carrion beetles and the skin beetles. In the United States, types of carrion beetles seen on the body may include *Thanatophilus lapponicous*, *Necrophila americana*, *Oiceoptoma rugulosum* and the burying beetle *Nicrophorus tomentosus*. Beetles usually make up the most of the population of insects on the body during black putrefaction. This beetle mass can vary in the types of families, making it hard to designate a certain beetle family to this stage of decay. Parasitic wasps continue to prey on these maggots and many generations of mites feed on the fluids let out by the body.

Butyric Fermentation

Butyric fermentation is the name of the fourth stage of decomposition. This fermentation process usually starts around 20–25 days after death. The body has finished flattening out from the previous putrefaction stage and the flesh and fluids on the body are slowly drying up. Butyric acid produces a distinct smell which is a component in breaking down the fluids in the body. This progression attracts different species of fauna to the carrion. Maggots and other insects that feed on soft flesh are unable to feed due to the drying out of the body; beetles and other insects with similar chewing mouth appendages are able to crush and chew the tougher segments of the dead body. At this time, most of the beetles are in the larval stage. Other insects such as cheese skippers and some parasitic wasps are also present. Hide Beetles from the family Trogidae and Carcass Beetles from the family Dermestidae are among the last beetles and generally the most common beetles seen during this fermentation period. The Hide Beetles as well as the Carcass Beetles are not predacious and are found on the tougher portions of the body such as bone and ligaments. Also, they are the only beetles that are capable of using an enzyme to break down proteins such as keratin. The cheese fly from the family Piophilidae, is attracted to the smell produced by butyric acid.

Dry Decay

The final stage in animal decomposition is dry decay. Dry decay begins between 25 and 50 days after death and can last up to a year. The only remnants of the body are dry skin, hair, and bones. Mummification occurs when dehydrated tissues lack the nutritional value to be broken down by other means. This is observed in dry heat or low humidity environments and can last for decades. The bones go through a process termed diagenesis that changes the organic to inorganic constituent ratio within the bones. Any odor exuding from the body at this point is merely the natural flora and fauna associated with the area.

The fauna seen at this stage is limited. Bacteria feed on the hair and skin of the body, attracting many mites. Certain tineid moths also feed on the remaining hair. Silphidae, a family of carrion beetles, may still be present during this stage. They typically arrive early in decomposition and stay until dry decay feeding on the larvae of other insects. Beetles of the family Nitidulidae can also be seen inhabiting the body. The normal soil fauna of the environment will begin to return during this stage.

Factors Affecting Decomposition

Understanding how a corpse decomposes and the factors that may alter the rate of decay is extremely important for evidence in death investigations. Campobasso, Vella, and Introna consider the factors that may inhibit or favour the colonization of insects to be vitally important when determining the time of insect colonization.

Temperature and Climate

Low temperatures generally slow down the activity of blow-flies and their colonization of a body. Higher temperatures in the summer favour large maggot masses on the carrion. Dry and windy environments can dehydrate a corpse, leading to mummification. Dryness causes cessation in bacterial growth since there are no nutrients present to feed on.

Access

Access to the body can limit which insects can get to the body in order to feed and lay eggs. In the United States, corpses found in brightly lit areas are generally inhabited by *Lucilia illustris*. This is in contrast to *Phormia regina*, which prefers more shaded areas. Darkness, cold, and rain limit the amount of insects that would otherwise colonize the body. A submerged corpse can vary in temperature and is colonized by very few terrestrial insects. Fish, crustaceans, aquatic insects and bacteria would be the likely fauna in this case. Bodies that have been buried are harder to get to than freely available bodies which limits the availability of certain insects to colonize. The Coffin fly *Megaselia scalaris* is one of the few fly species seen on buried bodies because it has the ability to dig up to six feet underground to reach a body and oviposit.

Reduction and Cause of Death

Scavengers and carnivores such as wolves, dogs, cats, beetles, and other insects feeding on the remains of a carcass can make determining the time of insect colonization much harder. This is because the decomposition process has been interrupted by factors that may speed up decomposition. Corpses with open wounds, whether pre or post mortem, tend to decompose faster due to easier insect access. The cause of death likewise can leave openings in the body that allow insects and bacteria access to the inside body cavities in earlier stages of decay. Flies oviposit eggs inside natural openings and wounds that may become exaggerated when the eggs hatch and the larvae begin feeding.

Clothing and Pesticides

Wraps, garments, and clothing have shown to effect the rate of decomposition because the corpse is covered by some type of barrier. Wraps, such as tight fighting tarps can advance the stages of decay during warm weather when the body is outside. However, loose fitting coverings that are open on the ends may aid colonization of certain insect species and keep the insects protected from the outside environment. This boost in colonization can lead to faster decomposition. Clothing also provides a protective barrier between the body and insects that can delay stages of decomposition. For instance, if a corpse is wearing a heavy jacket, this can slow down decomposition in that particular area and insects will colonize elsewhere. Bodies that are covered in pesticides or in an area surrounded in pesticides may be slow to have insect colonization. The absence of insects feeding on the body would slow down the rate of decomposition.

Percent Body Fat of Corpse

More fat on the body allows for faster decomposition. This is due to the composition of fat, which is high in water content. Larger corpses with higher percent body fat also tend to retain heat much longer than corpses with less body fat. Higher temperatures favour the reproduction of bacteria inside high nutrient areas of the liver and other organs.

Drugs

On occasion, drugs that are present in the body at death can also affect how fast insects break down the corpse. Development of these insects can be sped up by cocaine and slowed down by drugs containing arsenic.

Current Research

New research in the related field entomotoxicology is currently studying the effects of drugs on the development of insects who have fed on the decomposing tissue of a drug user. The effects of drugs and toxins on insect development are proving to be an important factor when determining the insect colonization time. It has been shown that cocaine use can accelerate the development of maggots. In one case, *Lucilia sericata* larvae that fed in the nasal cavity of a cocaine abuser, grew over 8 mm longer than larvae of the same generation found elsewhere on the body. Other researchers in entomotoxicology are developing techniques to detect and measure drug levels in older fly pupae. This research is useful for determining cause of death on bodies that are found during later stages of decay. To this date, bromazepam, levomepromazine, malathion, phenobarbital, trazolam, oxazepam, alimemazine, clomipramine, morphine, mercury, and copper have been recovered from maggots.

Conclusion

Understanding the stages of decomposition, the colonization of insects, and factors that may affect decomposition and colonization are key in determining forensically important information about the body. Different insects colonize the body throughout the stages of decomposition. These stages are categorized into fresh, putrefaction, black putrefaction, butyric fermentation, and dry decay. Identifying which insect is present can be useful in determining what stage of decay the body is in. Some insects are only found in certain regions and this information can help determine if a body has been moved by the type of maggots feeding on the body.

The environment surrounding the body is also very important to consider when using insect evidence. Insects, such as the coffin fly *Megaselia scalaris*, are nearly always found on bodies that have been buried. Using evidence that is very species specific can reveal a great deal of information about a body.

Forensic Entomology

This site was created in order to assist in the education of crime scene technicians, homicide investigators, coroners, medical examiners, and others involved in the death investigation process. A basic knowledge of the general appearance of insect of forensic importance, and a basic understanding of the proper methods for their collection, will allow investigators to make accurate and representative collections from the death scene. Enhanced knowledge on the behalf of law enforcement officials will provide for better communication between police, medical examiners, coroners, and forensic entomologists. Hopefully, improved communication will promote more frequent use of entomological resources, and more accurate collection of entomological evidence. This will allow forensic entomologists to be more precise in their statements and determinations, particularly in the area of postmortem interval estimations. Those involved in crime scene analysis and the death investigation process are encouraged to use this site as a reference resource to enhance their investigation efforts. It is impossible to determine when insect evidence my prove useful in a death investigation. Therefore, all law enforcement agencies should assemble a "Entomological Evidence Collection Kit" so that they are prepared in the event entomological evidence is encountered during any crime scene investigation.

What is Forensic Entomology?

Forensic Entomology is the use of the insects, and their arthropod relatives that inhabit decomposing remains to aid legal investigations. The broad field of forensic entomology is commonly broken down into three general areas: medicolegal, urban, and stored product pests. The medicolegal section focuses on the criminal component of the legal system and deals with the necrophagous (or carrion) feeding insects that typically infest human remains. The urban aspect deals with the insects that affect man and his immediate environment. This area has both criminal and civil components as urban pests may feed on both the living and the dead. The damage caused by their mandibles (or mouthparts) as they feed can produce markings and wounds on the skin that may be misinterpreted as prior abuse. Urban pests are of great economic importance and the forensic entomologist may become involved in civil proceedings over monetary damages. Lastly, stored product insects are commonly found in foodstuffs and the forensic entomologist may serve as an expert witness during both criminal and civil proceedings involving food contamination.

How Diverse is Forensic Entomology?

Here are a few examples: The diverse applications of forensic entomology include the detection of abuse in children and neglect of the elderly. Published cases exist that detail parents intentionally using wasps and bees to sting their children as a form of punishment. Additionally, entomological evidence has been used to prove neglect and lack of proper care for wounds existing on the elderly under both private and institutional care.

It is theorized that the stings (or mere presence) of bees and wasps may be responsible for a large number of single occupant car accidents that seem to lack a definitive cause. Some accident studies have shown insects to be within the top 20 causes of automobile accidents. In addition to automobile accidents, insects have been suspected of causing aircraft crashes through the obstruction of essential instrumentation, and even implicated in the obstruction of fuel lines causing engine failure.

Forensic entomologists are also requested to examine the fragmented remains of insects that have impacted and lodged on the front fascia, windshield, and radiator of automobiles. Analysis of such remains can yield evidence to the probable path of an automobile through particular areas when pinpointing the location and areas of travel are of unique importance.

Insects can also affect the interpretation of blood spatter pattern analysis. Roaches simply walking through pooled and splattered blood will produce tracking that may not be readily recognizable to the untrained observer. Specks of blood in unique and unusual areas (such as on ceilings) may mislead crime scene technicians unless they are aware of the appearance of blood contaminated roach tracks. Similarly, flies and fleas may also track through pooled and spattered blood. However, flies will also feed on the blood and then pass the partially digested blood in its feces, which are known as "flyspecks". Flies will also regurgitate and possibly drop a blood droplet on a remote surface, which may serve to confuse bloodstain analysis.

Fleas feeding on the living pass a large amount of undigested blood (used as the larval food source) on many household surfaces. If a crime occurs in a heavily infected apartment, fecal drops already present would serve to confuse analysts as those droplets would test positive for human blood. Therefore it is important to recognize and properly document the natural artifacts that may occur from the presence, feeding, and defecation of roaches, flies, and fleas. Insects that feed on living, decomposing, or dried vegetable material are submitted to the forensic entomologist in an effort to determine the country or point of origin. This is particularly important with vegetative material such as imported cannabis.

What information can a forensic entomologist provide at the death scene?

Forensic entomologists are commonly called upon to determine the postmortem interval or "time since death" in homicide investigations. More specifically, the forensic entomologist estimates a portion of the postmortem interval based on the age of the insect present. This entomological based estimation is most commonly called the "Time Since Colonization". Based on the factors in a particular investigation, this may, or may not, closely approximate the entire postmortem interval. In either case, it is the duty of the Forensic Pathologist, Medical Examiner, or Coroner to estimate the postmortem interval, and the Forensic Entomologist may assist them in providing information on the "time since colonization", which can ultimately be used to substantiate a portion of the postmortem interval.

The forensic entomologist can use a number of different techniques including species succession, larval weight, larval length, and a more technical method known as the accumulated degree hour technique which can be very precise if the necessary data is available. A qualified forensic entomologist can also make inferences as to possible postmortem movement of a corpse. Some flies prefer specific habitats such as a distinct preference for laying their eggs in an outdoor or indoor environment. Flies can also exhibit preferences for carcasses in shade or sunlit conditions of the outdoor environment. Therefore, a corpse that is recovered indoors with the eggs or larvae of flies that typically inhabit sunny outdoor locations would indicate that someone returned to the scene of the crime to move and attempt to conceal the body.

Similarly, freezing or wrapping of the body may be indicated by an altered species succession of insects on the body. Anything that may have prevented the insects from laying eggs in their normal time frame will alter both the sequence of species and their typical colonization time. This alteration of the normal insect succession and fauna should be noticeable to the forensic entomologists if they are familiar with what would normally be recovered from a body in a particular environmental habitat or geographical location. The complete absence of insects would suggest clues as to the sequence of postmortem events as the body was probably either frozen, sealed in a tightly closed container, or buried very deeply.

Entomological evidence can also help determine the circumstances of abuse and rape. Victims that are incapacitated (bound, drugged, or otherwise helpless) often have associated fecal and urine soaked clothes or bed dressings. Such material will attract certain species of flies that otherwise would not be recovered. Their presence can yield many clues to both antemortem and postmortem circumstances of the crime.

Currently, it is now possible to use DNA technology not only to help determine insect species, but to recover and identify the blood meals taken by blood feeding insects. The DNA of human blood can be recovered from the digestive tract of an insect that has fed on an individual. The presence of their DNA within the insect can place suspects at a known location within a definable period of time and recovery of the victims' blood can also create a link between perpetrator and suspect.

The insects recovered from decomposing human remains can be a valuable tool for toxicological analysis. The voracious appetite of the insects on corpses can quickly skeletonize the remains. In a short period of time the fluids and soft tissues needed for toxicological analysis disappear. However, it is possible to recover the insect larvae and run standard toxicological analyses on them as you would human tissue. Toxicological analysis can be successful on insect larvae because their tissues assimilate drugs and toxins that accumulated in human tissue prior to death.

Entomological Collection Equipment

Proper equipment is essential for the proper collection of entomological evidence. Preconfigured Forensic Entomology Kits are commercially available. However, due to budget constraints some police departments are unable to purchase ready-made collection kits and are forced to seek a low cost alternative. It is possible to configure your own collection kit, and most of the necessary items can be purchased at discount stores. A list of necessary items has been compiled to assist in the compilation of your own collection kit. The more "hard-to-find" speciality items such as insect nets and vials are discussed below:

Nets: Aerial insect nets are an essential component of a forensic entomology collection kit. The most popular and economical net has an 18" wood handle and sturdy 12" diameter galvanized steel ring. The net should be constructed from polyester and have a reinforcing muslin ring band. These nets can be purchased from most biological and forestry supply companies.

Vials: Collection vials can be purchased from about any biological supply house. "Screw-cap" types are best, and the typical size is a 4 dram vial with a foil insert in the lid. This vial is suitable for most crime scene collections and is commonly used by law enforcement officials. However, this vial is not suitable for long-term or archival storage of specimens since the preservation fluid will evaporate at a slow rate and periodic maintenance is required.

For long term storage (for archival or museum purposes), vials are available containing a neoprene cone insert in the lid, which reduces evaporation to negligible levels and periodic maintenance is not required as often.

Forceps: The proper forceps are useful for the collection of entomological specimens without damage. Soft-touch or "Feather-touch" forceps are best as they are constructed of thin and flexible metal that prevents the larval body from being crushed if too much force is applied by the collector. However, a cheaper and very adequate forceps are commonly available at discount stores. Biological and forestry supply houses commonly stock a wide selection of speciality forceps.

Collection Containers

As part of the proper processing of crime scenes for entomological evidence, living larvae must be collected and shipped to the forensic entomologist for rearing purposes. These larvae must be kept on a food source during shipment. The best method for ensuring their survival is to construct a container to house them during shipment. This container should be ridged in construction and have a tight fitting lid. It must also be large enough to contain a foil pouch holding with a palm sized piece of beef or chicken liver and approximately 50-60 larvae. Recommended containers to temporarily house the larvae for transport and shipping are plastic 16oz. cups. However, bait containers, and drink cups (with lids) from fast food restaurants are also used and will suffice if nothing else is available. Styrofoam cups can also be used, but they are not suitable for housing beetle larvae.

Insect

Insects (Class Insecta) are arthropods, having a hard exoskeleton, a three-part body (head, thorax, and abdomen), three pairs of jointed legs, compound eyes, and two antennae. They are the most diverse group of animals on the planet and include approximately 30 gladiator and icebug, 35 *Zoraptera*, 150 snakefly, 200 silverfish, 300 alderfly, 300 webspinner, 350 jumping bristletail, 550 scorpionfly, 600 *Strepsiptera*, 1,200 caddisfly, 1,700 stonefly, 1,800 earwig, 2,000 flea, 2,200 mantis, 2,500 mayfly, 3,000 louse, 3,000 walking stick, 4,000 cockroach, 4,000 lacewing, 4,000 termite, 5,000 dragonfly, 5,000 thrips, 5,500 booklouse, 20,000 cricket, grasshopper, and locust, 82,000 true bug, 110,000 ant, bee, sawfly, and wasp, 120,000 true fly, 170,000 butterfly and moth, and 360,000 beetle species described to date. The number of extant species is estimated at between six and ten million , with over a million species already described. Insects represent more than half of all known living organisms and potentially represent over 90% of the differing life forms on Earth. Insects may be found in nearly all environments, although only a small number of species occur in the oceans, a habitat dominated by another arthropod group, the crustaceans.

Adult modern insects range in size from a 0.139 mm (0.00547 in) fairyfly (*Dicopomorpha echmepterygis*) to a 56.7-centimetre (22.3 in) long stick insect (*Phobaeticus chani*). The heaviest documented present-day insect was 70 g (2½ oz) Giant Weta, though the Goliath beetles *Goliathus goliatus*, *Goliathus regius* and Cerambycid beetles such as *Titanus giganteus* hold the title for some of the largest species in general.

The largest known extinct insect is a kind of dragonfly, *Meganeura*.

Body Structure

Insects possess segmented bodies supported by an exoskeleton, a hard outer covering made mostly of chitin. The segments of the body are organized into three distinctive but interconnected units, or tagmata; a head, a thorax, and an abdomen. The head supports a pair of sensory antennae, a pair of compound eyes, if present, one to three simple eyes or (ocelli) and three sets of variously modified appendages that form the mouthparts. The thorax has six segmented legs (one pair each for the prothorax, mesothorax and the metathorax segments making up the thorax) and two or four wings (if present in the species). The abdomen (made up of eleven segments some of which may be reduced or fused) has most of the digestive, respiratory, excretory and reproductive internal structures.

Nervous System

The nervous system of an insect can be divided into a brain and a ventral nerve cord. The head capsule (made up of six fused segments) has six pairs of ganglia. The first three pairs are fused into the brain, while the three following pairs are fused into a structure called the subesophageal ganglion.

The thoracic segments have one ganglion on each side, which are connected into a pair, one pair per segment. This arrangement is also seen in the abdomen but only in the first eight segments. Many species of insects have reduced numbers of ganglia due to fusion or reduction. Some cockroaches have just six ganglia in the abdomen, whereas the wasp *Vespa crabro* has only two in the thorax and three in the abdomen. And some, like the house fly *Musca domestica*, have all the body ganglia fused into a single large thoracic ganglion.

Until very recently, no one had ever documented the presence of nociceptors (the cells that detect and transmit sensations of pain) in insects, though recent findings of nociception in larval fruit flies challenges this and raises the possibility that some insects may be capable of feeling pain.

Digestive System

An insect uses its digestive system to extract nutrients and other substances from the food it consumes. Most of this food is ingested in the form of macromolecules and other complex substances (such as proteins, polysaccharides, fats, and nucleic acids) which must be broken down by catabolic reactions into smaller molecules (i.e. amino acids, simple sugars, etc.) before being used by cells of the body for energy, growth, or reproduction. This break-down process is known as digestion.

The insect's digestive system is a closed system, with one long enclosed tube called the alimentary canal which runs lengthwise through the body. The alimentary canal only allows food to enter the mouth, and then gets processed as it travels toward the anus. Each of the three sections of the alimentary canal performs a different process of digestion. In addition to the alimentary canal, insects also have paired salivary glands and salivary reservoirs. These structures usually reside in the thorax.

The salivary glands (30) produce saliva, the salivary ducts lead from the glands to the reservoirs and then forward through the head to an opening called the salivarium behind the hypopharynx; which movements of the mouthparts help mix saliva with food in the buccal cavity. Saliva mixes with food which travels through salivary tubes into the mouth, beginning the process of breaking it down.

The first section of the alimentary canal is the fore-gut (27) or stomodaeum. In the fore-gut, initial breakdown of large food particles occurs, mostly by saliva. The fore-gut includes the Buccal cavity, the esophagus, and the crop, which stores food before it passes to the mid-gut.

Once food leaves the crop, it passes to the mid-gut (13) or mesenteron. The mid-gut is where digestion really happens, through enzymatic action. Microscopic projections from the mid-gut wall, called microvilli, increase surface area and allow for maximum absorption of nutrients.

In the hind-gut (16) or proctodaeum, is where undigested food particles join uric acid from Malphigian tubules to form fecal pellets. The rectum absorbs most of the water in this waste matter, and the dry pellet is then eliminated through the anus (17), there by completing the process of digestion.

Respiration and Circulation

Insect respiration is accomplished without lungs using a system of internal tubes and sacs through which gases either diffuse or are actively pumped, delivering oxygen directly to tissues that need oxygen. Since oxygen is delivered directly, the circulatory system is not used to carry oxygen, and is therefore greatly reduced; it has no closed vessels, consisting of little more than a single, perforated dorsal tube which pulses peristaltically, and in doing so helps circulate the hemolymph inside the body cavity. Air is taken in through spiracles, openings on the sides of the abdomen. There are many different patterns of gas exchange demonstrated by different groups of insects. Gas exchange patterns in insects can range from continuous, diffusive ventilation, to discontinuous gas exchange.

Scanning Electron Micrograph of A Thrips (Thysanoptera), Showing Fine Structure, The Compound Eyes, Wing Construction, and Setae.

Exoskeleton

Most higher insects have two pairs of wings located on the second and third thoracic segments. Insects are the only invertebrates to have developed flight capability, and this has played an important part in their success. Insect flight is not very well understood, relying heavily on turbulent aerodynamic effects. The primitive insect groups use muscles that act directly on the wing structure. The more advanced groups making up the Neoptera have foldable wings and their muscles act on the thorax wall and power the wings indirectly. These muscles are able to contract multiple times for each single nerve impulse, allowing the wings to beat faster than would ordinarily be possible.

Their outer skeleton, the cuticle, is made up of two layers; the epicuticle which is a thin and waxy water resistant outer layer and contains no chitin, and another layer under it called the procuticle. This is chitinous and much thicker than the epicuticle and has two layers, the outer being the exocuticle while the inner is the endocuticle. The tough and flexible endocuticle is built from numerous layers of fibrous chitin and proteins, criss-crossing each others in a sandwich pattern, while the exocuticle is rigid and sclerotized. The exocuticle is greatly reduced in many soft-bodied insects, especially the larval stages.

Reproduction

Most insects hatch from eggs, but some are ovoviviparous or viviparous, and all undergo a series of moults as they develop and grow in size. This manner of growth is necessitated by the inelastic exoskeleton. Moulting is a process by which the individual escapes the confines of the exoskeleton in order to increase in size, then grows a new and larger outer covering. In some insects, the young are called nymphs and are similar in form to the adult except that the wings are not developed until the adult stage. This is called *incomplete metamorphosis* and insects showing this are termed *hemimetabolous*. *Holometabolous* insects show *complete metamorphosis*, which distinguishes the Endopterygota and includes many of the most successful insect groups. In these species, an egg hatches to produce a larva, which is generally worm-like in form, and can be divided into five different forms; eruciform (caterpillar-like), scarabaeiform (grublike), campodeiform (elongated, flattened, and active), elateriform (wireworm-like) and vermiform (maggot-like). The larva grows and eventually becomes a pupa, a stage marked by reduced movement and often sealed within a cocoon.

There are three types of pupae; obtect (the pupa is compact with the legs and other appendages enclosed), exarate (where the pupa has the legs and other appendages free and extended) and coarctate (where the pupa develops inside the larval skin). In the pupal stage, the insect undergoes considerable change in form to emerge as an adult, or imago. Butterflies are an example of an insect that undergoes complete metamorphosis. Some insects have even evolved hypermetamorphosis.

Some insects (parasitic wasps) show polyembryony where a single fertilized egg can divide into many and in some cases thousands of separate embryos. Other developmental and reproductive variations include haplodiploidy, polymorphism, paedomorphosis, sexual dimorphism, parthenogenesis and more rarely hermaphroditism.

Senses and Communication

Many insects possess very sensitive and/or specialized organs of perception. Some insects such as bees can perceive ultraviolet wavelengths, or detect polarized light, while the antennae of male moths can detect the pheromones of female moths over distances of many kilometres. There is a pronounced tendency for there to be a trade-off between visual acuity and chemical or tactile acuity, such that most insects with well-developed eyes have reduced or simple antennae, and vice-versa. There are a variety of different mechanisms by which insects perceive sound, and it is by no means universal; the general pattern, however, is that if an insect can produce sound, then it can also hear sound, though the range of frequencies they can hear is often quite narrow (and may in fact be limited to only the frequency that they themselves produce).

Some nocturnal moths can perceive the ultrasonic emissions of bats, a mechanism which helps them avoid predation. Certain predatory and parasitic insects can detect the characteristic sounds made by their prey/hosts. Bloodsucking insects have special sensory structures that can detect infrared emissions, and use them to home in on their hosts.

A few such insects also have a well-developed number sense, such as the solitary wasps that provision with a single species of prey. The mother wasp lays her eggs in individual cells and provides each egg with a number of live caterpillars on which the young feed when hatched. Some species of wasp always provide five, others twelve, and others as high as twenty-four caterpillars per cell. The number of caterpillars is different among species, but it is always the same for each sex of larvae. The male solitary wasp in the genus *Eumenes* is smaller than the female, so the mother of one species supplies him with only five caterpillars; the larger female receives ten caterpillars in her cell. She can in other words distinguish between both the numbers five and ten in the caterpillars she is providing and which cell contains a male or a female.

Light Production and Vision

A few insects, such as members of the families Poduridae and Onychiuridae, Mycetophilidae, and the beetle families Lampyridae, Phengodidae, Elateridae and Staphylinidae are bioluminescent. The most familiar group are the fireflies, beetles of the family Lampyridae. Some species are able to control this light generation to produce flashes. The function varies with some species using them to attract mates, while others use them to lure prey. Cave dwelling larvae of *Arachnocampa* glow to lure small flying insects into sticky strands of silk. Some fireflies of the genus *Photuris* mimic the flashing of female *Photinus* species to attract males of that species, which are then captured and devoured. The colours of emitted light vary from dull blue to the familiar greens and the rare reds.

Most insects except some species of cave dwelling crickets are able to perceive light and dark. Many species have acute vision capable of detecting minute movements. The eyes include simple eyes or ocelli as well as compound eyes of varying sizes. Many species are able to detect light in the infrared, ultraviolet as well as the visible light wavelengths. Colour vision has been demonstrated in many species and phylogenetic analysis suggests that the basic bauplan of UV-green-blue trichromacy existed from at least the Devonian period.

Sound Production and Hearing

Insects were the earliest organisms to produce sounds and to sense them. Soundmaking in insects is achieved mostly by mechanical action of appendages. In the grasshoppers and crickets this is achieved by stridulation. The cicadas have the loudest sounds among the insects and have special modifications to their body and musculature to produce and amplify sounds. Some species such as the African cicada, *Brevisana brevis* have been measured at 106.7 decibels at a distance of 50 cm (20 in). Some insects, such as the hawk moths and Hedylid butterflies, can hear ultrasound and take evasive action when they sense detection by bats. Some moths produce ultrasound clicks that were earlier thought to have a role in jamming the bat echolocation, but it was subsequently found that these are produced mostly by unpalatable moths to warn bats, just as warning colourations are used against predators that hunt by sight. These calls are also made by other moths involved in mimicry.

Very low sounds are also produced in various species of Neuroptera, Lepidoptera (butterflies and moths), Coleoptera and Hymenoptera produced by the mechanical actions of movement often aided by special microscopic stridulatory structures. Most sound-making insects also have tympanal organs that can perceive airborne sounds. Most insects are also able to sense vibrations transmitted by the substrate. Communication using substrate-borne vibrational signals is more widespread among insects because of the size constraints in producing air-borne sounds. Insects cannot effectively produce low-frequency sounds, and high-frequency sounds tend to disperse more in a dense environment (such as foliage), so insects living in such environments communicate primarily using substrate-borne vibrations. The mechanisms of production of vibrational signals are just as diverse as those for producing sound in insects.

Some species use vibrations for communicating within members of the same species, such as to attract mates as in the songs of the shield bug *Nezara viridula* while it can also be used to communicate between entirely different species, such as between ants and myrmecophilous lycaenid caterpillars. The Madagascar hissing cockroach has the ability to press air through the spiracles to make a hissing noise, and the Death's-head Hawkmoth makes a squeaking noise by forcing air out of their pharynx.

Chemical Communication

In addition to the use of sound for communication, a wide range of insects have evolved chemical means for communication. These chemicals, termed semiochemicals, are often derived from plant metabolites include those meant to attract, repel and provide other kinds of information.

While some chemicals are targeted at individuals of the same species, others are used for communication across species. The use of scents is especially well known to have developed in social insects.

Social Behaviour

Social insects, such as termites, ants and many bees and wasps, are the most familiar species of eusocial animal. They live together in large well-organized colonies that may be so tightly integrated and genetically similar that the colonies of some species are sometimes considered superorganisms.

It is sometimes argued that the various species of honey bee are the only invertebrates (and indeed one of the few non-human groups) to have evolved a system of abstract symbolic communication, where a behaviour is used to *represent* and convey specific information about something in the environment), called the dance language-the angle at which a bee dances represents a direction relative to the sun, and the length of the dance represents the distance to be flown.

Only those insects which live in nests or colonies demonstrate any true capacity for fine-scale spatial orientation or homing-this can be quite sophisticated, however, and allow an insect to return unerringly to a single hole a few millimetres in diameter among a mass of thousands of apparently identical holes all clustered together, after a trip of up to several kilometres' distance, and (in cases where an insect hibernates) as long as a year after last viewing the area (a phenomenon known as philopatry).

A few insects migrate, but this is a larger-scale form of navigation, and often involves only large, general regions.

Care of Young

Most insects lead short lives as adults, and rarely interact with one another except to mate or compete for mates. A small number exhibit some form of parental care, where they will at least guard their eggs, and sometimes continue guarding their offspring until adulthood, and possibly even actively feeding them. Another simple form of parental care is to construct a nest (a burrow or an actual construction, either of which may be simple or complex), store provisions in it, and lay an egg upon those provisions. The adult does not contact the growing offspring, but it nonetheless does provide food. This sort of care is typical of bees and various types of wasps.

Locomotion

Flight

Insects are the only group of invertebrates to have developed flight. The evolution of insect wings has been a subject of debate. Some proponents suggest that the wings are para-notal in origin while others have suggested they are modified gills. In the Carboniferous age, some of the *Meganeura* dragonflies had as much as a 50 cm (20 in) wide wingspan.

The appearance of gigantic insects has been found to be consistent with high atmospheric oxygen. The percentage of oxygen in the atmosphere found from ice core-samples was as high as 35% compared to the current 21%. The respiratory system of insects constrains their size, however the high oxygen in the atmosphere allowed larger sizes.

The largest flying insects today are much smaller and include several moth species such as the Atlas moth and the White Witch (*Thysania agrippina*). Insect flight has been a topic of great interest in aerodynamics due partly to the inability of steady-state theories to explain the lift generated by the tiny wings of insects.

Unlike birds, insects are swept along by the prevailing winds. This includes aphids, which are often transported long distances by low-level jet streams. As such, fine line patterns associated with converging winds within weather radar imagery, like the WSR-88D, are dominated by insect returns.

Walking

Many adult insects use six legs for walking and have adopted a tripedal gait. The tripedal gait allows for rapid walking whilst always having a stable stance and has been studied extensively in cockroaches. The legs are used in alternate triangles touching the ground.

For the first step the middle right leg and the front and rear left legs are in contact with the ground and move the insect forward, whilst the front and rear right leg and the middle left leg are lifted and moved forward to a new position. When they touch the ground to form a new stable triangle the other legs can be lifted and brought forward in turn and so on.

The purest form of the tripedal gait is seen in insects moving at speed. However, this type of locomotion is not rigid and insects can adapt a variety of gaits; for example, when moving slowly, turning, or avoiding obstacles, four or more feet may be touching the ground. Insects can also adapt their gait to cope with the loss of one or more limbs.

Cockroaches are amongst the fastest insect runners and at full speed actually adopt a bipedal run to reach a high velocity in proportion to their body size. As cockroaches move extremely rapidly, they need recording at several hundred frames per second to reveal their gait.

More sedate locomotion is also studied by scientists in stick insects Phasmatodea. A few insects have evolved to walk on the surface of the water, especially the bugs of the family, Gerridae, also known as water striders. A few species of ocean-skaters in the genus *Halobates* even live on the surface of open oceans, a habitat that has few insect species. Insect walking is of particular interest as an alternative form of locomotion to the use of wheels for robots (Robot locomotion).

Swimming

A large number of insects live either parts or the whole of their lives underwater. In many of the more primitive orders the immature stages are aquatic while some other groups have aquatic adults as well.

Many of these species have adaptations to help in locomotion under water. The water beetles and water bugs have legs adapted into paddle like structures. Dragonfly naiads use jet propulsion, forcibly expelling water out of the rectal chamber. Some species like the water striders are capable of walking on the surface of water.

They can do this because their claws are not at the tips of the legs as in most insects, but recessed in a special groove further up the leg; this prevents the claws from piercing the water's surface film. Other insects such as the Rove beetle *Stenus* are known to emit salivary secretions that reduce surface tension making it possible for them to move on the surface of water by Marangoni propulsion (also known by the German term *Entspannungsschwimmen*).

Species that are submerged also have adaptations to aid in respiration. Many larval forms have gills that can extract oxygen dissolved in water, while others need to rise to the water surface to replenish air supplies which may be held or trapped in special structures.

Evolution

The relationships of insects to other animal groups remain unclear. Although more traditionally grouped with millipedes and centipedes, evidence has emerged favouring closer evolutionary ties with the crustaceans. In the Pancrustacea theory, insects, together with Remipedia and Malacostraca, make up a natural clade. Other terrestrial arthropods, such as centipedes, millipedes, scorpions and spiders, are sometimes confused with insects since their body plans can appear similar, sharing (as do all arthropods) a jointed exoskeleton. However upon closer examination their features differ significantly; most noticeably they do not have the six legs characteristic of adult insects.

The higher level phylogeny of the arthropods continues to be a matter of debate and research. In 2008, a Tufts University student and a faculty lecturer uncovered what they believe is the world's oldest known full-body impression of a primitive flying insect, a 300 million-year-old specimen from the Carboniferous Period.

The scientists presented the fossil at the Second International Congress on Ichnology, in Krakow, Poland in September 2008. The oldest definitive insect fossil is the Devonian *Rhyniognatha hirsti*, from the 396 million year old Rhynie chert. This species already possessed dicondylic mandibles, a feature associated with winged insects, suggesting that wings may already have evolved at this time. Thus, the first insects probably appeared earlier, in the Silurian period.

The origins of insect flight remain obscure, since the earliest winged insects currently known appear to have been capable fliers. Some extinct insects had an additional pair of winglets attaching to the first segment of the thorax, for a total of three pairs. So far, there is no evidence that suggests that the insects were a particularly successful group of animals before they got their wings.

Late Carboniferous and Early Permian insect orders include both several current very long-lived groups and a number of Paleozoic forms. During this era, some giant dragonfly-like forms reached wingspans of 55 to 70 cm, (22-28 in) making them far larger than any living insect. This gigantism may have been due to higher atmospheric oxygen levels that allowed increased respiratory efficiency relative to today.

The lack of flying vertebrates could have been another factor. Most extinct orders of insects developed during the Permian era that began around 270 million years ago. Many of the early groups became extinct during the Permian-Triassic extinction event, the largest mass extinction in the history of the Earth, around 252 million years ago.

The remarkably successful Hymenopterans appeared in the Cretaceous but achieved their diversity more recently, in the Cenozoic. A number of highly-successful insect groups evolved in conjunction with flowering plants, a powerful illustration of co-evolution.

Many modern insect genera developed during the Cenozoic; insects from this period on are often found preserved in amber, often in perfect condition. Such specimens are easily compared with modern species. The study of fossilized insects is called paleoentomology.

Coevolution

Insects were among the earliest terrestrial herbivores and acted as major selection agents on plants. Plants evolved chemical defences against this herbivory and the insects in turn evolved mechanisms to deal with plant toxins. Many insects make use of these toxins to protect themselves from their predators. Such insects often advertise their toxicity using warning colours. This successful evolutionary pattern has also been utilized by mimics. Over time, this has led to complex groups of co-evolved species. Conversely, some interactions between plants and insects are beneficial, and coevolution has led to the development of very specific mutualisms in such systems.

Classification

Traditional morphology-based systematics has included in the subphylum Hexapoda four groups-Insects (Ectognatha), springtails (Collembola), Protura and Diplura, the latter three being grouped together as Entognatha on the basis of internalized mouthparts. Supraordinal relationships have undergone numerous changes with the advent of cladistic methods and genetic data. A recent hypothesis is that Hexapoda is polyphyletic, with the entognath classes having separate evolutionary histories from Insecta.

As many of the traditional morphology-based taxa have been shown to be paraphyletic, it is best to avoid using terms such as subclass, superorder and infraorder and instead focus on monophyletic groupings. The following list represents the best supported monophyletic groupings for the Insecta.

Neoptera

Polyneoptera

- Caloneurodea †
- Titanoptera †
- Protorthoptera †
- Plecoptera (stoneflies)
- Embiidina=Embioptera (webspinners)
- Zoraptera (angel insects)
- Dermaptera (earwigs)
- Orthoptera (grasshoppers, crickets, katydids)
- Phasmatodea (stick insects)
- Notoptera (ice-crawlers & gladiators)
- Blattaria (cockroaches)
- Isoptera (termites-included in Blattaria)
- Mantodea (mantids).

Paraneoptera

- Psocodea (booklice, barklice)
- Mallophaga (chewing lice-in Psocodea)
- Anoplura (sucking lice-in Psocodea)

- Thysanoptera (thrips)
- Hemiptera (true bugs, aphids, cicadas).

Endopterygota= Holometabola

- Glosselytrodea †
- Miomoptera †
- Hymenoptera (wasps, bees, ants)
- Coleoptera (beetles)
- Strepsiptera (twisted-winged parasites).

Neuropterida

- Raphidioptera (snakeflies)
- Megaloptera (alderflies, dobsonflies)
- Neuroptera (lacewings, antlions).

Antliophora/Mecopteroidea

- Mecoptera (scorpionflies, hangingflies)
- Siphonaptera (fleas-in Mecoptera)
- Diptera (true flies)
- Protodiptera †.

Amphiesmenoptera

- Trichoptera (caddisflies)
- Lepidoptera (butterflies, moths, skippers).

Insects can be divided into two groups, historically treated as subclasses: Apterygota (wingless) and Pterygota (winged). The Apterygota consists of two primitively wingless orders-Archaeognatha (bristletails) and Thysanura (silverfish).

Archaeognatha makes up the Monocondylia (based on mandibular morphology) while Thysanura and Pterygota are grouped together as Dicondylia. It is possible that the Thysanura itself is not monophyletic, with the family Lepidotrichidae a sister group to the Dicondylia (Pterygota and the remaining Thysanura).

Paleoptera and Neoptera are the winged orders of insects, separated by the presence of sclerites and musculature that allow for folding of the wings flat over the abdomen in the latter group. Neoptera can further be divided into hemimetabolous and Holometabolous groups.

It has proven particularly difficult to elucidate interordinal relationships within Polyneoptera. Phasmatodea and Embiidina have been suggested to form Eukinolabia. Mantodea, Blattodea & Isoptera are thought to form a monophyletic group termed Dictyoptera. Paraneoptera has turned out to be more closely related to Endopterygota than to the rest of the Exopterygota. The recent molecular finding that the traditional louse orders Mallophaga and Anoplura are derived from within Psocoptera has led to the new taxon Psocodea.

It is quite likely that Exopterygota is paraphyletic in regards to Endopterygota. Contentious matters include Strepsiptera and Diptera grouped together as Halteria based on a reduction of one of the wing pairs-a position not well-supported in the entomological community.

The Neuropterida are often lumped or split on the whims of the taxonomist. Fleas are now thought to be closely related to boreid mecopterans. Many questions remain to be answered when it comes to basal relationships amongst endopterygote orders, particularly Hymenoptera.

The study of the classification or taxonomy of any insect is called systemic entomology. Normally, if one chooses to work with a more specific order or even a family, the sytemics would be added to the study of that order or family, an example would be systemic dipterology.

Relationship to Humans

Many insects are considered pests by humans. Insects commonly regarded as pests include those that are parasitic (mosquitoes, lice, bed bugs), transmit diseases, damage structures (termites), or destroy agricultural goods (locusts, weevils). Many entomologists are involved in various forms of pest control, often using insecticides, but more and more relying on methods of biocontrol.

Although pest insects attract the most attention, many insects are beneficial to the environment and to humans. Some pollinate flowering plants (for example wasps, bees, butterflies, and ants). Pollination is a trade between plants that need to reproduce, and pollinators that receive rewards of nectar and pollen. A serious environmental problem today is the decline of populations of pollinator insects, and a number of species of insects are now cultured primarily for pollination management in order to have sufficient pollinators in the field, orchard or greenhouse at bloom time.

Insects also produce useful substances such as honey, wax, lacquer and silk. Honey bees have been cultured by humans for thousands of years for honey, although contracting for crop pollination is becoming more significant for beekeepers.

The silkworm has greatly affected human history, as silk-driven trade established relationships between China and the rest of the world. In some cultures, insects, especially deep-fried cicadas, are considered to be delicacies, and, in fact, have a high protein content for their mass.

In most first-world countries, however, the consumption of insects is eschewed, although peoples in these cultures tend to accidentally consume between 50 and

90 insects in a given year, though this number varies based on region. Fly larvae (maggots) were formerly used to treat wounds to prevent or stop gangrene, as they would only consume dead flesh. This treatment is finding modern usage in some hospitals. Adult insects such as crickets, and insect larvae of various kinds are also commonly used as fishing bait.

In some parts of the world, insects are used for human food, while being a taboo in other places. There are proponents of developing this use to provide a major source of protein in human nutrition. Since it is impossible to entirely eliminate pest insects from the human food chain, insects already are present in many foods, especially grains. Most people do not realize that food safety laws in many countries do not prohibit insect parts in food, but rather limit the quantity. According to cultural materialist anthropologist Marvin Harris, the eating of insects is taboo in cultures that have protein sources that require less work, like farm birds or cattle.

Many insects, especially beetles, are scavengers, feeding on dead animals and fallen trees, recycling the biological materials into forms found useful by other organisms, and insects are responsible for much of the process by which topsoil is created. The ancient Egyptian religion adored dung beetles and represented them as beetle-shaped amulets, or scarabs.

The most useful of all insects are insectivores, those that feed on other insects. Many insects can potentially reproduce so quickly that if all of their offspring were to survive, they could literally bury the earth in a single season. However, for any given insect, there will be plenty of species of insects that are either parasitoids or predators that play a significant role in controlling it. This role in ecology is usually assumed to be primarily one of birds, but insects, though less glamorous, are much more significant.

Human attempts to control pests by insecticides can backfire, because important but unrecognised insects already helping to control pest populations are also killed by the poison, leading eventually to population explosions of the pest species.

4

Anatomy of Insect

Introduction

Because of the great diversity of form exhibited by the insects any introduction to anatomy like this is only going to be able to cover the basics, within each order and family these familiar themes are replayed in a myriad of different ways creating what might seem to be a bewildering array of different body plans. A closer look will reveal the same basic plan in all insects, at least in the adult forms. Below are described some of the basic parts that make up an insects body.

An insect body has a hard exoskeleton protecting a soft interior, it is divided into three main parts (the head, thorax and abdomen) each of which is in turn composed of several smaller segments.

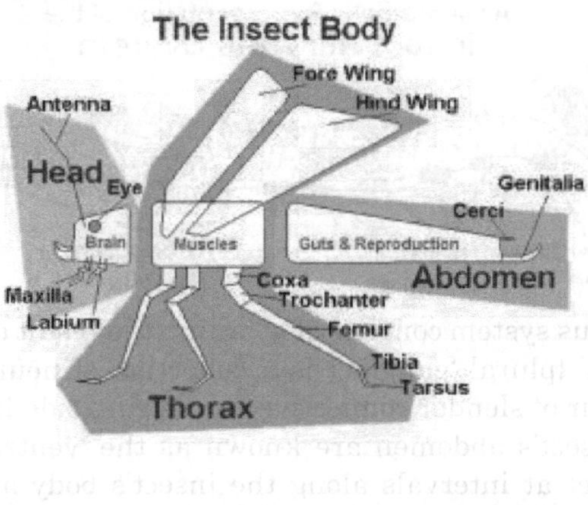

The Integument

The amazing success of the insects must in part lie with the incredible mixture of flexibility and strength of the integument (that is the part of an insect that makes up the hard exoskeleton) that allows insects their freedom of movement without loss of defence and protection. It is made up of three parts, the most visible of which is the outer 'cuticle' and its attendant bristles and hairs, below this are the 'epidermis' and the 'basement membrane'

The Cuticle

The cuticle is a relatively thin layer of non-cellular material which lines the external surface of the body as well as lining the tracheae the anterior and posterior sections of the alimentary canal and parts of the reproductive system. It is flexible, elastic and white when first formed and stays this way in many larval forms, however in most adults it undergoes chemical processes which result in hardening and darkening and which are referred to as 'sclerotization'.

The cuticle can be divided into two layers, a very thin outer layer called the epicuticle which contains no chitin and is highly resistant to water and other solvents.

Beneath this is the much thicker procuticle which can again be divided into two distinct layers, an outer 'exocuticle' which lies immediately below the 'epicuticle' and an inner 'endocuticle' which consists of a large number of layers of protein and chitin fibres laid down in a laminated pattern such that the individual strands in each layer cross each other thus creating an extremely tough and flexible substance. Below the cuticle lie the other two components of the integument, the 'epidermis' which is a single layer of secretary cells and the 'basement membrane' which is an amorphous layer about 0.5 micrometres thick.

The Nervous System

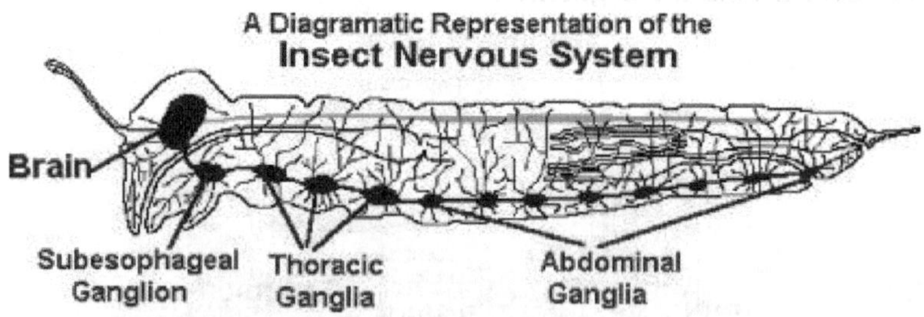

The insect nervous system consists of a 'brain' (the result of the fusion of 3 pairs 'ganglia' [a 'ganglion' {plural 'ganglia'} is a collection of neurons or nerve cells in a single place]). A pair of slender connectives cords run, side-by-side from the brain to the end of the insect's abdomen are known as the 'ventral nerve cord'. These connective cords meet at intervals along the insect's body at the 'ganglia'.

In the most primitive design there is one pair of ganglia per body segment, thus, as the head is made up out of 6 fused body segments, it contains 6 pairs of ganglia, these are collected into 2 groups, each of 3 ganglia, the foremost of which is called the brain and the hindmost the 'subesophageal ganglion'. The ganglia function to coordinate the activities of the body segment they represent.

In the most basic design there are usually 3 thoracic ganglia and 8 abdominal ganglia but in most of the higher insects some of abdominal ganglia have been lost, or become fused with those nearer the head. In the cockroach *Blatta orientalis* there are 3 thoracic and only 6 abdominal ganglia and in the hornet *Vespa crabro* there are only 2 thoracic ganglia and 3 abdominal ganglia, the yhind most of which is very mich larger than the other two because it is made up of a number of ganglia fused together.

This trend culminates in insects such as the common house fly, *Musca domestica*, where all the abdominal and thoracic ganglia have become fused into a single compound body-ganglion.

Inside the Insect

Breathing

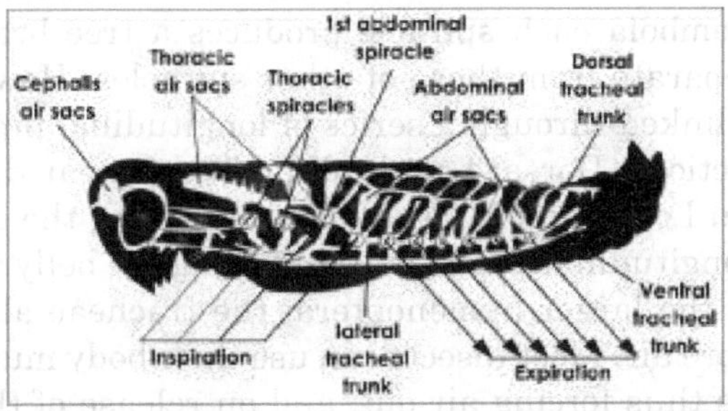

Insects have no lungs! Most insects breath passively through their 'Spiracles' (special openings in the side of there cuticle) and the air reaches the body by means of a series of smaller and smaller pipes called 'Tracheae' when their diametre is large and 'Tracheoles' when their diametre is very small. Diffusion of gases is effective over small distances but not over larger ones, this is one of the reasons insects are all relatively small. Insects which do not have spiracles and tracheae, such as some Collembola, breath directly through their skins, also by diffusion of gases.

The number of spiracles an insect has is variable between species, however they always come in pairs, one on each side of the body, and usually one per segment. Some of the Diplura have eleven pairs, with four pairs on the thorax, but in most of the ancient forms of insects, such as Dragonflies and Grasshoppers there are two thoracic and eight abdominal spiracles.

However in most of the remaining insects there are less; so that Hoverflies, Syrphidae, have only two pairs, both of which are on the thorax and none on the abdomen while many Mosquito larvae and aquatic Beetle larvae have only one abdominal pair of spiracles. Many insects have valves that allow them to close their spiracles, thus preventing water loss.

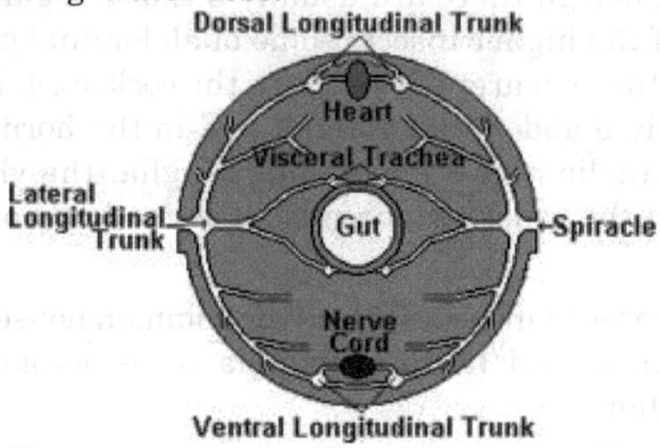

Diagramatic Representation in TS of the Insect Tracheal System.

The tracheae, which are thin pipes, spread out from the spiracles to reach the whole body, with the smallest tracheoles contacting single muscle cells individually. In some of the Collembola each spiracle produces a tree branch, or tree root, of tracheae that are separate from those of other spiracles. However in most insects the tracheae are all linked through a series of longitudinal pipes called trunks and many smaller connections. Dorsal Longitudinal Trunk near the top, or back, of the insect's body; Lateral Longitudinal Trunk running along the sides just in from the spiracles; Ventral Longitudinal Trunk running along the belly of the insect. In many insects, particularly the larger hymenoptera, the tracheae also link to a series of airsacs which can store air. Most insects can use their body muscles to squeeze their tracheae and airsacs thus forcing air out, and on release of the muscular tension, drawing fresh air into the large tracheae. In insects like the dragonflies this is a continuous action, but in others it is irregular, as in cockroaches, or only occurs after active exercise as in the larger hymenoptera.

Hearing and Touch

Ears

Many but not all insects can hear sounds, some even hear sounds that we can't hear ourselves. Insects hear through one of four different ways, the most common of which is the tympanum. Tympanal organs always occur as paired organs, they are composed of a thin cuticular membrane stretched across an air space of some sort and some form of connection to the nervous system. In the Orthoptera tympanum are common, though situated in different places in different species, i.e. on the first thoracic segment in Grasshoppers and on the front legs in the Crickets. Tympanal organs also occur in the Cicada and some families of the Lepidoptera,. The other three forms of hearing organs are 1) Johnston's Organ, via the movement of hairs on the antennal scape i.e. the Mosquito *Aedes aegypti*. 2) Auditory Hairs these occur on some Lepidopteran larvae as well as on some Orthoptera. 3) The Pilifer; this is a unique auditory organ found only in the head of certain species of Hawk Moths of the subfamily Choerocampinae its optimum frequency is between 30 and 70 kHz which would allow it to hear the echolocation calls of many of the larger insectivorous bats.

Touch

Touch is an extremely important sense to insects and like smell insects have developed many different ways to detect mechanical stimulus, these all involve some form of physical change in the receptor, the most common are hairs attached to nerves which react when the hairs are moved, these are called Trichoid sensilla. Another common type looks more like a drum with something pressing up against the skin of the drum from beneath, these are called Campaniform sensilla. Mechanoreceptors detect not only the physical interaction with another body but also air movements, changes in air pressure and also changes in the stresses being applied to the insects cuticle, thus allowing it to better control its movements and maintain balance. Insects also use modified forms of the various sensory detectors described above to detect, changes in temperature, humidity and also in some cases to detect infra red radiation, x-ray radiation and the Earth's magnetic field.

The Insect Head

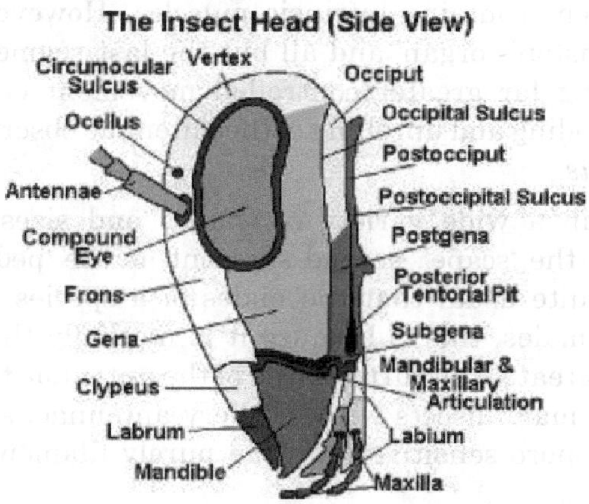

The insect's head is sometimes referred to as the head-capsule, and is the insect's feeding and sensory centre. It supports the eyes, antennae and and jaws of the insect. The upper-mid portion of an insects face is called the 'frons' below this is the 'clypeus' and below this the 'labrum' to either side of which may be seen the edges of the 'mandibles' in some insects various aspects of the 'maxilliary' palps may extend beyond and or below these even when viewed from front on.

The 'frons' = that area of the face below the top two 'ocelli' and above the 'frontoclypeal sulcus' (if and when this is visible) and in between the two 'frontogenal sulci', it supports the 'pharyngeal dilator' muscles and in immature forms it bears the lower two arms of the ecdysial cleavage lines.

The 'clypeus' = that area of the face immediately below the frons (with which it may be fused in the absence of the frontoclypeal sulcus) and the frontoclypeal sulcus. It supports the 'cibarial dilator' muscles and may be divided horizontally into a 'post.' and 'anteclypeus'. The 'labrum' = is equivalent to the insect's upper lip and is generally moveable, it articulates with the clypeus by means of the 'clypeolabral suture'. The rest of the front of the head: that bit which is above the frons is known as the 'vertex'; the sides of the head are known as the 'gena'.

The Antennae

The antennae are an insects primary, non-visual, sense organs, though in a few rare instances they have become adapted for other purposes such as seizing prey items (i.e. the larva of Chaoborus sp. {Diptera}) or holding females during mating (i.e. the males of Meloe sp. {Coleoptera})

Not all insects possess antennae, they are absent from the Protura.

In most insects the antennae possesses a mechanosensory organ on the pedicel (the second antennal segment) called 'Johnston's organ' and, normally, only the basal antennal segment contains intrinsic muscles. However in two orders the antennae lack a 'Johnston's organ' and all but the last segment contains intrinsic muscles, thus allowing far greater controlled movement of the antennae as is demonstrated by the rolling and unrolling of the antennae observed in the Collembola *Tomocerus longicornus.*

Antennae come in a wide variety of shapes and sizes, generally the first segment is known as the 'scape', second segment as the 'pedicel' and the rest as the flagellum. It is quite usual that the males of a species have more elaborate antennae than the females, this is because it is normally the males who have to find the females. The greater the surface area of the antennae the more dilute scents they can detect, thus male insects with feathery antennae, such as those seen in many moths, are far more sensitive than the purely filamentous ones of crickets and cockroaches.

The Eyes

Though some species of insects have been shown to be able respond to light stimulus through their cuticle, most light sensitivity occurs through one or more eyes. Insects possess two different sorts of eyes, the usually large and obviously visible compound eyes, and two varieties of ocelli or simple eyes.

Compound Eyes

Compound eyes are so named because the cornea is composed of a number of individual facets or lenses, rather than a single lens as in ocelli. The number of separate visual elements or ommatidia varies greatly between species as well as between the larger taxa, so that while worker ants of different species may have between 1 and 600 ommatidia per single eye, adult male Odonata may have more than 28,000 per single eye.

This creates a considerable difference in the presentation of light stimulus to the insect brain, however the ability of insects to navigate the world by means of visual stimuli suggest that they have overcome the problems inherent in this multi-faceted perception. The physical differences between single-lens and multi-lens perception are excellently shown at the Bee Keepers Home Page in the section on "the world through the eye of a bee".

Much like our eyes, the eyes of insects, can be divided into four basic parts: the supportative material that keeps all the parts together; a light gathering part (the lens and the auxilary lens called a 'crystalline cone'); a light receptor that converts the recieved light into electrical energy; and the nerves that carry the electrical impulses to the brain for analysis. In the compound eyes of insects these parts are repeated numerous times side by side in a space saving hexagonal pattern.

Transverse Section Through An Insect Eye

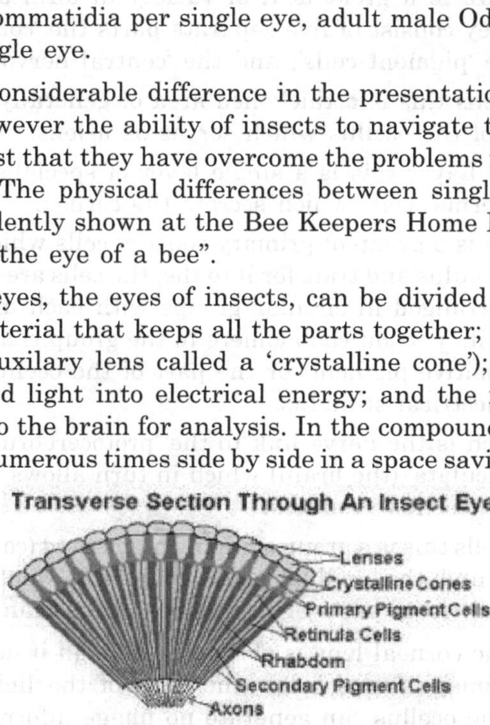

The lens is formed by a transparent and colourless cuticle and it is usually biconvex. Beneath this is the crystalline cone (which is comprised of four cells called 'Semper cells' after the man who first described them). Normally this functions as a secondary lens.

The receptive parts of an insect's eye are the 'retinula cells'. Each ommatidium normally has eight retinula cells arranged to leave a central core space in the centre of the ommatidium, into which each retinula cell projects a series of microvilli (like very small fingers). These microvilli are the actual light detecting part of the cells and are collectively referred to as the rhabdomere. The eight (or occasionally 7 or 9) rhabdomeres form a rhabdom.

The corneal lens is supported by 'primary pigment cells' and the retinula cells and associated rhabdoms are supported by 'secondary pigment cells'. The retinula cells are connected to axons at the base of the eye, it is these which carry the information collected by the lenses and converted into electrical impulses by the rhabdom to the brain, thus allowing the insect to see.

Simple Eyes

Ocelli are present in most insects to some degree, though as with all aspects of insect anatomy there is a great deal of variety in form and even in relative function. Generally they consist of five separate parts the 'cornea', the 'corneagen layer', the 'retina', the 'pigment cells', and the 'central nervous connections'.

1) The Corneal Lens this is a thickened area of generally transparent cuticle to the outside of the ocellus which serves as a lens.
2) The Corneagen Layer this is a single layer of specialised transparent and colourless epidermal cells which secrete the cornea.
3) The Retina this is a group of primary sensory cells which convert light into an electrical stimulus and transfer it to the; the cells are called 'retinula' cells and they are arranged in circular groups with each member of the group contributing to its portion rhabdomere to the group rhabdom. The rhabdom is the light sensitive pigment, or the part of the ocellus that converts the light into an electrical stimulus.
4) The Axon, which is the nerve link to the 'protocerebrum' and hence to the 'Corpora pedunculata' (the brain) which in turn allows the insect to use the information the ocellus produces.
5) The Pigment Cells this is a group of highly pigmented (coloured) cells variably distributed around the ocellus whose main roll would appear to be the exclusion of light from parts of the ocellus other than the cornea.

The function of the corneal lens is obscure, although it does project an image into the ocellus this image forms below the level of the light-sensitive cells, or rhabdom. Therefore the ocellus can generate no image information, however it is very sensitive to low levels of light and to changes in light intensity and scientists believe that the ocelli are useful in allowing the insect to detect the horizon, to respond quickly to changes in light intensity. Two different forms of ocelli have been described for insects, Dorsal ocelli and Lateral ocelli.

Dorsal ocelli occur mostly in adult insects and are situated on the front of the insects face in the area of the 'frons' and or the 'epicranium', lateral ocelli generally occur on the sides of the insect head and are the form of eye most common in larval forms; there are a number of concrete differences between the two forms which can be found explained in any competent entomological text book such as Imm's 1984.

Smell

Most insects communicate using smell or chemoreception and it is not surprising that they have evolved a large variety of ways of detecting the molecules involved.

Insects do not have noses like us which concentrate all our sense of smell in one place, instead they have a lot of small sensory bodies scattered over their body, though they tend to have a concentration of them on their antennae. We can recognise several different common forms of chemoreceptor though these are not the only forms they can take by any means,

1) Sensilla trichoidea, hair-like structures commonly found on the feet of flies and the antennae of many insects, they are the most common form of chemoreceptor found.
2) Sensilla basiconica and Sensilla styloconica, these are peg-like or cone-like and are thicker and more solid than trichoid sensilla, these are commonly found on the antennae, though they also occur on the maxillary palps of Lepidopteran larvae and the ovipositor of the Blowfly *Phormia regina.*
3) Sensilla coeloconica or pit-peg organs these are always situated in a pit as their common name suggests, unlike the previous two which project above the insects cuticle. They are common on a variety of insect antennae and in *Apis mellifera* they detect Carbon Dioxide.
4) Sensilla placodea these differ from the first three in that they consist of a flat plate of cuticle, they occur on the antennae of various Aphids and *Apis mellifera.*

The Mouth Parts

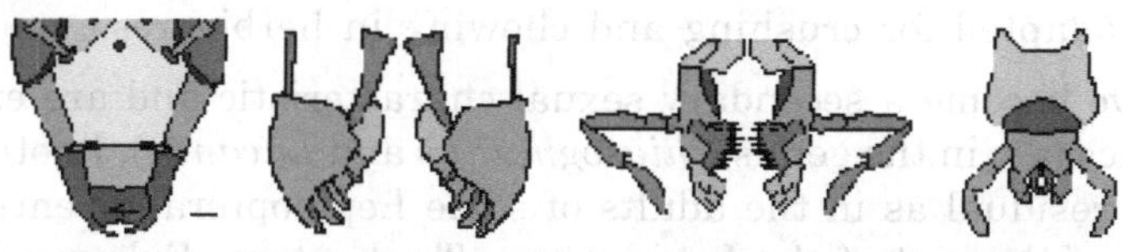

In the primitive form the insect mouth articulates (moves) from side to side in a horizontal plane, rather than vertically as do ours. In those groups of insects that evolved later the basic mouthparts shown below have become highly modified, in the butterflies and moths they are transformed into a long flexible straw, in the blood and sap sucking insects of various orders they have become a hypodermic needle and in most of the flies you see from day to day they evolved into a extensible and highly effective sponge. However in all these cases, all, or most of the basic parts are still there, see the tranverse section of the needle like mouth of the mosquito right as an example. The Insect mouth consists of a number of parts which, starting from the foremost, are called 1) The Labrum, 2) The Mandibles, 3) The maxillae, 4) The Labium, 5) The Hypopharynx or tongue. The Labrum is normally a simply structure that is equivalent to the insect's upper lip and is generally moveable, it articulates with the clypeus by means of the 'clypeolabral suture'.

The Mandibles

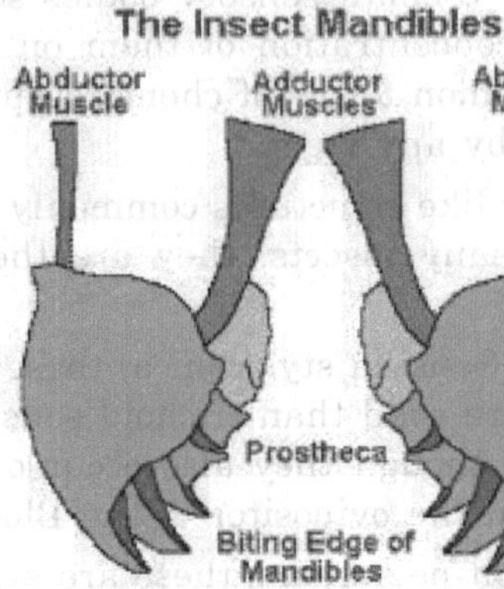

The mandibles, with the maxillae, the labial palps and in some species the hypopharynx constitute the moveable aspects of the insect mouth, the mandibles and the maxillae are the equivalent of jaws with the exception that they move transversely.

The mandibles show great variety within the insect orders and like our more familiar teeth they are hard and show variation in accordance with diet, thus they are sharp edged in carnivores, being extremely sickle like in the ant *Aceton burchelli*, whilst being adapted for crushing and chewing in herbivores.

They have become a secondary sexual characteristic and are extremely large in some Beetles (i.e. in the genus *Chiasognathus* and *Lucanus*). In other orders they have become residual as in the adults of some Lepidoptera, or entirely absent as in the adults of the rest of the Lepidoptera, Trichoptera, Ephemeroptera and the Diptera.

Mandibles are used not only for feeding but also for attack and defence, becoming extremely exaggerated in various species of termites ants, and for manipulation of materials as in the nest building insects particularly the Hymenopteran Bees, Ants and Wasps.

The Maxillae

The maxillae are a pair of modified limbs which work behind the mandibles and in front of the labium as a pair of accessory jaws. They are composed of the following parts.

1) The Cardo, this is the piece nearest the head capsule and in some species of insect it is the only part of the maxillae that is connected to the head.
2) The Stypes this is central bulk of the maxillae and supports the;

The Insect Maxillae (one side only)

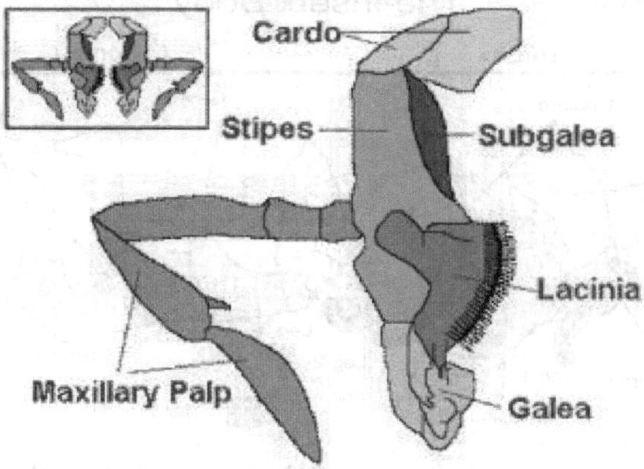

3) Palpifer, which in turn supports the;
4) Maxilliary palp, which has one to seven segments and is mainly used as a sensory organ.
5) The Lacinia is situated at the distal end of the Stypes, is often serrated or toothed and serves to aid the eating process both by holding and masticating the food. It is boarded distally (to the far side in relation to the overall body) by;
6) The Galea and proximally (to the near-side in relationship to the body).

The Labium

The Labium results from the fusion of a pair of limbs and serves a purpose similar to our lower lip for the insects.

The main body of the labium is divided into three parts the central 'mentum' which is boarded on either side by the 'submentum' proximally which hinges with the head, and the prementum distally.

The prementum supports two pairs of lobes known as the 'glossae' and to the outside of them the 'paraglossae', and a pair of labial palps which are primarily sensory in function. The glossae and paraglossae may be fused, with one or the other considerably reduced, in which case the whole thing is known as the 'ligula'.

The hyper-pharynx or tongue is found behind the mouth, and has the salivary ducts at its base, in most Diptera (true flies) and Hemiptera (true bugs) it has become highly modified and serves as the main feeding organ, in many cases combining with the rest of the mouthparts to form a stylette or piercing organ.

The Insect Thorax

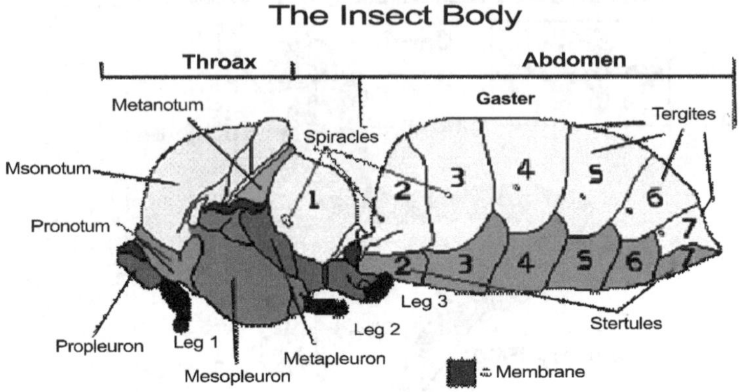

The thorax is the main engine room of the insect. It like the Abdomen is built up of a series of concave upper and convex lower integumental plates, in the Thorax they are known as the 'Nota' and 'Pleura' the whole being held together by a tough yet stretchable membrane. The thorax can be conveniently divided into three separate and normally easily visible sections called from the front, the 'prothorax' the 'mesothorax' and the 'metathorax'. This leads to the nota and pleura being named 'Pronotum', 'Mesonotum', 'Metanotum' and 'Propleuron', 'Mesopleuron' and 'Metapleuron'.

Within the whole class insecta the form and shape of the thorax shows considerable variation depending on the needs of each species. In species that are strong fliers, for example, the Mesothorax may become greatly enlarged.

In the adult insects, as well as in the nymphs of those insects with an 'incomplete metamorphosis' (those whose life cycle lacks the production of a pupa), each of these sections supports a pair of legs. In the pterygote (winged) insect orders the first two sections support a pair of wings, except in the Diptera (true flies) where the second or hind wings have become modified into a pair of club-like balancing organs called 'halteres'.

The Legs

The typical insect leg as shown in consists of six main sections;

1) The Coxa, this is the most basal aspect of the insect leg and articulates with the 'sternites'.
2) The Trochanter is usually small and serves as a joint between the 'coxa' and the 'femur'.
3) The Femur is usually long and stouter than the other segments and contains the main muscles used in running, jumping and digging.

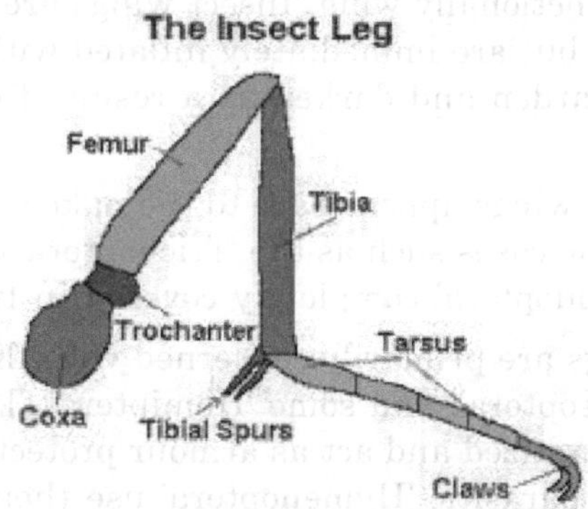

4) The tibia is also generally long serving to increase the length of the leg, as well as adding an extra joint and thus extra flexibility the underside of the tarsal segments may possess pads.
5) The Tarsus is the foot of the insect leg and can consist of between one and five segments.
6) The Claws are situated at the end of the 'tarsus' and serve to assist the insect in holding onto the substrate or to its prey. Between the claws may be found a special pad known as an 'arolium' and which acts using suction developed by large numbers of minute tubular hairs to help hold the insect to smooth substrates.

As in all aspects of insect anatomy their is an amazing diversity of form to be found in the insect leg, the three examples shown below are colour coded to the same pattern as the labelled leg above. Can you remember which part is which?

The Wings

Unlike legs, wings occur only in the adult forms of those insects which possess them, in those insects with 'incomplete metamorphosis' small wing buds are visible in the place the wings will occupy in the adult.

Generally speaking wings are thin flat structures consisting of two fine membranes supported by a series of sclerotized veins. In many small insects the veins may be absent and in some very small insects, such as the Thrips, which are only one or two millimetres long, the wings consist of a central midrib supporting a series of fine hairs, these look more like feathers.

Wings are not totally stiff but bend and flex in an amazing manner during flight greatly improving their aerodynamics.

Most species of winged insect have two pairs of wings and in many of the more advanced orders these two wings are held together by a variety of mechanisms to form a single, larger, functionally wing. Insect wings are soft and shapeless when the adults first emerge, but are immediately inflated with blood pressure through the veins before they harden and darken as a result of coming into contact with the air.

Though most insect wings appear bare to the naked eye many are dotted with minute hairs and in some cases such as the 'Trichoptera' completely clothed in fine hairs, or as in the 'Lepidoptera' completely covered in tiny scales.

Though insect wings are primarily concerned with flight they serve a number of purposes, in the 'Coleoptera', and some 'Hemiptera' (True Bugs) the fore wings have become highly sclerotized and act as armour protecting the insect concerned. While some very small parasitic 'Hymenoptera' use their wings to swim through the water to their hosts.

In Beetles the sclerotised forewings are called elytra, while the somewhat less stiffened forewings of Grasshoppers and Crickets are called tegmen. The bright colours of many insect wings are there for mating purposes and allow males and females to recognise each other, in others the colours perform a role in camouflage hiding the insect from its predators, and in others they perform a thermodynamic role, the different colours reflecting and absorbing different wave lengths of light allowing the insect to control its internal temperature. In some insects the wings may perform two or more of these functions simultaneously.

In two orders of insects one pair of wings has become completely reduced to small balancing organs and they provide no power of lift during flight at all. The most common of these are the Flies which have the hind wing reduced. Much less commonly seen but just as interesting are the Stylopids which have the forewing reduced.

In many insects the hind wings are reduced to some extent and thus smaller than the forewings, in others, the forewing has become stiffened to form a protective function the hind wing is often larger than the forewing, and therefore needs to be folded up when it is not in use. Grasshoppers and Crickets fold their wings longitudinally, while Beetles, and Earwigs the wings are folded both longitudinally and laterally. In Staphylinid betles and Earwigs the forewings are very short and the degree of folding of the hindwings is truly amazing.

The Insect Abdomen

The insect abdomen is built up of a series of concave upper integumental plates known as 'tergites' and convex lower integumental plates known as 'sternites', the whole being held together by a tough yet stretchable membrane. It contains the insects digestive tract and reproductive organs, it consists of eleven segments in most orders of insects though the 11th segment is absent in the adult of most the higher orders. In many of the Hymenoptera, and several other orders, the number of abdominal segments is reduced and in the Honey Bee only seven segments are visible. In the 'Collembola' the abdomen has only six segments. In the Hymenoptera there is a constriction where the 1st and 2nd abdominal segments meet, this is called the waste, and the remaining portion of the abdomen is called the gaster.

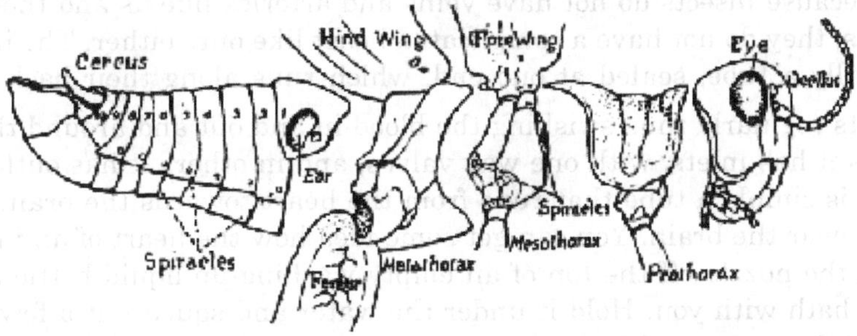

Unlike other Arthropods the insects possess no legs on the abdomen in the adult form, though the 'Protura' do have rudimentary leg-like appendages on the first three abdominal segments. Many larval insects particularly the 'Lepidoptera' and the 'Symphyta' have appendages called 'pseudo' or prolegs on their posterior abdominal segments as well as their more familiar thoracic legs these allow them to grip onto the edges of plant leaves as they walk around.

The Muscles

Like us Insects need muscles in order to move the various different bits of their bodies around, however insects have their muscles attached to the inside of their skeleton because like all the arthropods they have their skeletons on the outside of their body.

The inside of an insect's exoskeleton has special contours and bits and bobs on it which project inwards and allow for muscles to be attached and to help give them leverage, these projections are called 'apodemes'. The musculature of even the smallest insect can be as complicated as our own and makes for a fascinating study of design in miniature. The muscles of insects are generally light grey or translucent, unlike ours which appear red. This is because insects lack both the blood system that we have and the haemoglobin that makes our blood and hence our muscles red.

The Heart and the Blood

The haemolymph (blood) of insects flows freely around the inside of their bodies. Because an insect's haemolymph is not responsible for the transmission of oxygen to its cells, this is the job of the Trachaea, and therefore does not contain haemoglobin, it is not red. Normally it is a watery green colour, though it is pigmented (coloured) in some species. This haemolymph is a sort of soup rich in nutrients that flows around the inside of the insects body allowing the various organs to get at whatever resources they need and into which they dump their waste products.

These waste products are later removed from the haemolymph by the Malpighian tubules. Because insects do not have veins and arteries like us and the rest of the vertebrates, they do not have a complicated heart like ours either. The insect heart it is basically a tube, sealed at one end, which runs along their back.

It beats regularly thus swishing the blood in and out and around the body. In some cases it has inlets, with one way valves, and in others it has outlets as well. The Aorta is simple a tube that runs from the heart towards the brain, it is open at the end near the brain. You can get some idea how the heart of an insect works by cutting the nozzle off the top of an empty washing-up-liquid bottle and taking it into the bath with you. Hold it under the water and squeeze it a few times and watch how the water swirls around.

Here is a lovely drawing of the internal organs of an ant made by a talented Russian artist and physiologist called Nasonov before the great revolution in the late 1800s, it is very large and if you have a slow link you might like to see the smaller sectioned drawings below it instead though they have lost some clarity in resizing.

Digestion

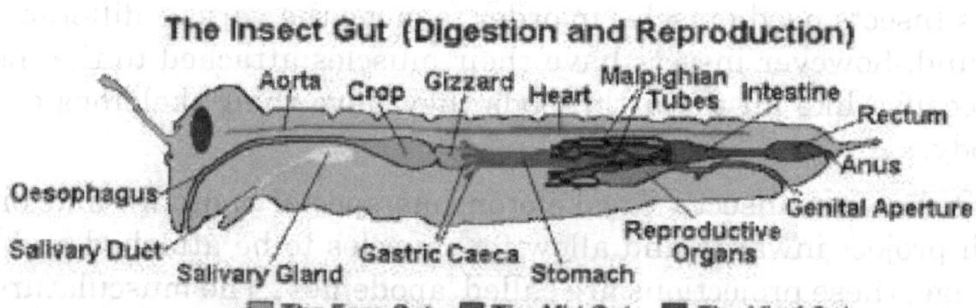

The Digestive system of an insect is usually a long straight tube running from the mouth to the anus, it is often divided into the 'fore gut', the 'mid gut' and the 'hind gut'. Immediately behind the mouth are the Salivary Glands, in most species these secrete saliva, generally a watery fluid that lubricates the food and contains a few enzymes to begin the processes of digestion.

However in some carnivorous insects the saliva is composed entirely of digestive enzymes, this applies particularly to those with external digestion of the food.

In other insects the salivary glands have become modified for purposes that have nothing to do with digestion. In Lepidopteran caterpillars and Caddisfly larvae they have been converted to the production of silk, while in the Queen Honey Bee they are called the mandibular glands and secrete hormones.

The Fore Gut

The fore gut is generally considered to consist of four sections, the Pharynx, the Oesophagus, the Crop and the Proventriculus. It is also known as the Stomodaeum.

The pharynx is the first part of the fore gut and apart from being a tube that connects the interior of the mouth area (sometimes known as the 'Buccal Cavity) with the more inward parts of the gut it sometimes serves as a pump to suck up the liquefied food of those insects which feed by means external digestion.

The Oesophagus is basically a tube leading to the mid gut via the crop and the proventriculus or gizzard. The crop is simply a storage area and the proventriculus, or gizzard, is a muscular extension of the crop. In those insects which feed on solid foods it is used to grind the food up into smaller particles, it can also serve as a filter to keep oversized particles out of the main digestive tract and as a valve controlling the flow of food into the midgut. The fore gut and the mid gut are separated by the 'stomodeal or cardiac valve'.

The Mid Gut

The midgut runs from the 'digestive or gastric caeca', a series of stubby pointed tubes leading off from the stomach to just before the Malpighian tubules, a series of long thin tubes. In between the two of these is the stomach, or ventriculus, which is the area of most active digestion. The gastric caeca serve to increase the surface area of the midgut, thus increasing both its ability to secrete digestive enzymes and its ability to extract useful products from the partially digested food.

The useful proteins, vitamins and fats that are released by the digestive processes pass across the wall of the midgut into the body cavity. The mid gut is lined by a semipermeable membrane composed of protein and chitin, like the cuticle, which allows the passage of liquids and dissolved substances to the midgut wall while preventing the passage of solid food particles, it is continually worn away by the passage of food through the gut and replaced by the epithelial cells of the mid gut wall. The mid gut and the hindgut are separated by the 'proctodeal valve'.

The Hind Gut

From the mid gut food passes to the hind gut (called the Proctodaeum in some books). The hind gut comprises the 'intestines' which is where much of the diffusion into the insects body occurs. The 'rectum' which compresses the undigested food and waste products, extracts more water from this if necessary before it is passed out through the 'anus' as faeces.

Arising from and the foremost part of the hind gut are the Malpighian tubules (named after Malpighi who discovered them) are not really to do with digestion at all but with elimination. They act like our kidneys and extract metabolic waste products (mostly nitrogenous ones such as urea, and uric acid) from the circulating body fluid called the haemolymph and excrete them into the intestines which is the first part of the hind gut.

Though insects possess a large number of digestive enzymes, they are often helped by the presence of symbiotic microorganisms, such as protozoa in the case of the termites and some primitive cockroaches which feed on wood, and bacteria in the wax moth *Galleria mellonella* which feeds on the wax that honey bees *Apis mellifera* uses to make the combs in its hives.

Reproductive Structures

Most insect species are bisexual, i.e. there are males and females in most species, these often look very different and have even been mistaken for different species in the past, some species are capable of reproduction without males, the eggs are unfertilised but develop and hatch into nymphs or larvae that are always female themselves, this is called 'parthenogenesis'.

Externally the sexual organs, called genitalia, of a female insect generally consist of an 'ovipositor' which is often encased in a pair of filaments called a 'sheath' and is which is used to by the female to put her eggs where she wants. Its form very greatly throughout the Insecta. The ovipositor of the Diptera is functionally similar i.e. it is used to lay eggs, but is morphologically distinct i.e. it arises or is made from different parts of the insects anatomy and should be called a 'pseudovipositor'. The median part of the oviduct which receives the aedeagus during mating is called the 'vagina'.

Externally the sexual organs of the male, also called genitalia, consist of a pair of 'claspers' which the male uses to hold onto the females genitalia and an intromittant organ called the 'aedeagus' which is the means by which the male passes the sperm onto the female.

Internally the female reproductive organs consist of a pair of ovaries which contain the ovarioles which is where the eggs or ova are formed, the bursa copulatrix which is where the sperm is first received) in those insects which have it) and a spermatheca which is where the sperm is stored. There are also various tubes down which the ova travel on their way from the ovaries to the outside world, fertilisation occurs in the common oviduct after the egg has received its shell or the 'chorion'.

To facilitate this the shell contains a very small opening at one end called the micropyle which allows the sperm to enter. As well as tubes there are several important glands some of which allow the female to keep the sperm alive and viable for a long time, as much as 20 years in some social insects; and some of which (collaterial glands) secrete the substances which allow the female to stick the eggs where she wants then to stay i.e. underneath a leaf, or to protect the eggs as in the ootheca produced by the Cockroaches and Mantids

Internally the male reproductive organs consist of a pair of 'testes' containing the 'testicular follicles' where the spermatozoa are made, the 'vas deferens' which is the tube down which the sperm travels, a 'seminal vesicle' which is where the sperm is stored prior to mating, and accessory glands which supply seminal fluid for additional volume and to nourish the sperm before and during their journey.

5

Commercial Greenhouse Poinsettia Production

History

Until the late 1980s, *T. vaporariorum* was the major pest of poinsettias. In 1986, an apparently new strain of *B. tabaci*, 'strain B,' caused substantial damage to poinsettias in Florida, and by 1991 had spread throughout the United States where it caused $500 million in damage. Prior to affecting poinsettias, *B. tabaci*, commonly called the tobacco, cotton, or sweet potato whitefly, was a common pest of agricultural crops. In 1994 *B. tabaci* 'strain B' was identified as a separate species, *B. argentifolii*, the silverleaf whitefly. The SLWF was given its name because when SLWF feed on the leaves of squash plants, characteristic silvering symptoms are produced. Another type of whitefly, the bandedwinged whitefly, is occasionally found on sticky traps in poinsettia production areas, but is seldom a problem on the crop. Although this publication focuses on the impact of GWF and SLWF on poinsettia crops, it is also important to note that both whiteflies are vectors for a variety of plant diseases. The SLWF is a vector of geminiviruses, which have been described as "some of the most devastating diseases of vegetables such as tomato, bean, and squash and of field crops such as tomato, beets, tobacco, and corn" (Agrios 1997).

Biology, Life Cycle, and Behaviour

Whiteflies land on the top surface of plant leaves and immediately walk around to the shaded lower side to feed and lay eggs. All life stages develop on the undersides of the leaves. The first instar of the nymph is called the crawler. The crawler emerges from the egg, moves a short distance, and begins to feed.

The developing whitefly remains immobile (sessile) for three more nymph instars then molts to become a mobile adult. In experiments with SLWF on a poinsettia crop, the timing of each stage at 72F is as follows:

- Average adult life-span 22.4 days

- Eggs laid per female 90.9 eggs
- Egg to adult emergence 49.9 days.

The rate of whitefly development is determined primarily by temperature, but host plant preferences play an important role. Experiments with GWF demonstrate that the rate of egg laying (oviposition), egg number laid per female, female longevity, total development time from egg to adult, and the mortality rates for all life stages are directly related to host plant nutrition. Moderate greenhouse temperatures (60–75F) favour the GWF, higher greenhouse temperatures (above 75F) favour the SLWF, but both thrive on poinsettia. If both the GWF and the SLWF are present in a poinsettia crop, the SLWF will out-compete and exclude the GWF in 50–60 days. The GWF and SLWF have wide host ranges: there are over 275 plant species affected by GWF and over 500 species affected by SLWF. Both the GWF and SLWF have a life cycle with four developmental stages—the egg, nymph, pupa and adult stage.

Identification

The SLWF and GWF can be distinguished in any life stage using the characteristics listed in Table 1.

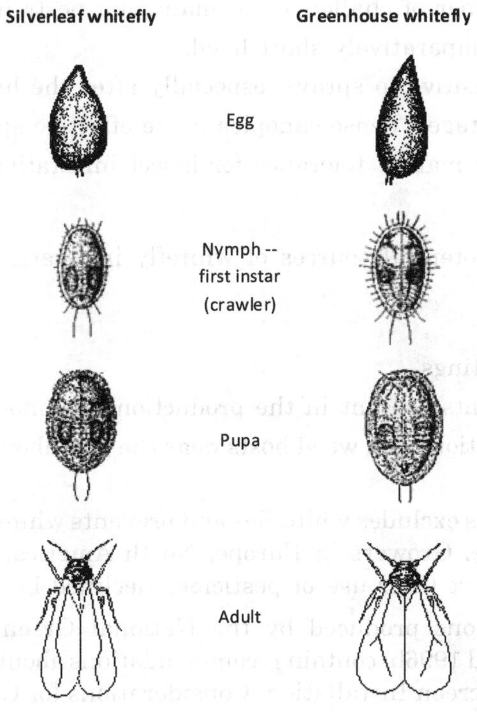

Figure 1. Life stages of the GWF and SLWF.

Damage

Whiteflies damage poinsettias in two ways; by direct feeding (chewing on the leaves) and by indirect feeding (tapping into the phloem to extract sugars). No matter which feeding method is used, relatively small populations of whiteflies make poinsettias unsalable.

Direct feeding damages poinsettias by causing bracts and stems to become chlorotic and bleached. Indirect feeding damages poinsettias when whiteflies use their piercing-sucking mouthparts to penetrate phloem, causing the whiteflies to excrete honeydew onto the leaves. Honeydew is sticky, full of undigested sugars, and is frequently invaded by sooty molds (nonparasitic fungi). The resultant "accumulations of sooty mold reduce the aesthetic quality and marketability of the crop, even though the plants are not directly injured".

The simple presence of whiteflies, even in the absence of other types of damage, reduces the aesthetic value of plants offered for retail sale.

Integrated Pest Management (IPM) of Whiteflies

There are a number of challenges in managing pests in poinsettias.

- The crop is comparatively short lived.
- Bracts are sensitive to sprays, especially after the bract begins to colour.
- In later crop stages, dense canopies make effective spraying more difficult.
- There is a low market tolerance for insect infestations.

Physical Controls

There are four potential sources of whitefly infestations in greenhouses, as follows:

- stock plants,
- purchased cuttings,
- other host plants present in the production greenhouse, and
- whitefly migration from weed hosts near the greenhouse or in the poinsettia crop itself.

Putting up screens excludes whiteflies and prevents whitefly migration. Screens are also cost effective. Growers in Europe, North America, and Israel who have installed screens report their use of pesticides declined by 50–90%.

Several publications produced by the National Greenhouse Manufacturers Association (1996a and 1996b) contain recommendations about screens. For example, "Greenhouse Insect Screen Installation: Considerations for Greenhouse Operators" provides a discussion of screen materials and construction methods, and "Standards for Ventilating and Cooling Greenhouse Structures" contains the necessary engineering formulas to compensate for the presence of screens. The North Carolina Commercial Flower Growers Association has published a detailed discussion of available screen materials, a comparison of their efficiency, and a list of screen manufacturers.

Table 1. Identifying whiteflies. Adapted from Sanderson 1995; Powell and Lindquist 1997; Gill and Sanderson 1998

	Silverleaf whitefly	Greenhouse whitefly
Eggs		
egg placement	underside of leaves, usually not in a pattern	underside of leaves in partial- or semi-circles
egg colour	turn brown-amber	turn grayish
Immatures		
nymphs (3 instars)	yellowish	greenish
pupae	rounded, few or no filaments on body perimeter	oval with filaments protruding from entire body perimeter
pupae, side view	sides curved or rounded	sides elevated
Adults		
size	smaller	
activity level	more active	
wing posture while on plants	45° toward vertical, like a tent	horizontal
body colour	white but occasionally with a yellowish hue	white

Examples of the calculations needed to expand the ventilation surface area to compensate for the reduced airflow through screens are available from the North Carolina Cooperative Extension Service; they are also available from the Web site listed above. Baker and Shearin have developed a PC-compatible program to guide you through the airflow/surface area calculations.

Cultural Controls

Good crop sanitation involves inspecting incoming poinsettia cuttings for the presence of whiteflies, followed by routine inspections (monitoring). Routine inspections allow you to identify whitefly infestations early and take appropriate corrective actions. All of the information in this "Cultural Controls" section is taken from Sanderson (1996).

A recommended practice is to thoroughly inspect, one month prior to the arrival of any new poinsettia cuttings, all the poinsettia plants currently growing in the greenhouse. If whiteflies are found, thoroughly spray the plants with insecticide to reduce the whitefly population, if possible, to zero. Three weeks later (one week prior to the arrival of the new cuttings), examine existing plants for the presence of whiteflies. If whiteflies are found, take appropriate control measures. Purchased cuttings should be carefully inspected *for all life stages* of the whitefly. Where possible, inspect newly purchased cuttings in a holding area separate from the poinsettia production areas. Sanderson recommends that "each shipment and cultivar should be inspected individually, because whitefly levels can differ by cultivar and propagator." Focus the inspection on the undersides of the three oldest (that is, lowest) leaves of the cuttings because lower leaves are more likely to harbour the immature life stages. Adult whiteflies prefer the upper leaves.

Monitoring

The best way to monitor greenhouses for the presence of whiteflies is to use yellow sticky cards. The *minimum* number of cards recommended for successful monitoring is one yellow sticky card per 1,000 square feet of greenhouse floor space, but one card per 250 square feet is more effective.

Use of sticky cards in poinsettia was pioneered by the New York State Poinsettia IPM Program in 1989–92. The program combines sampling adults using yellow sticky cards (3x5 inch) with leaf sampling to detect immature stages. Key features of the New York State Program include.

- organizing the production areas into areas containing 2,000 pots to create pest management units (PMUs),
- determining acceptable (that is, threshold) numbers of whitefly nymphs per leaf,
- scouting each PMU at a minimum of once a week for whiteflies,
- sampling each PMU in sequence and using action thresholds to decide whether pest control actions are needed, and
- tagging infested "sentinel plants" for follow-up inspection to determine if control measures are effective or if immature life stages are continuing to develop toward adulthood.

Sequential sampling for the New York State Poinsettia IPM Plan. Each PMU is sampled weekly, and control measures, if necessary, are determined for each PMU. There are three thresholds based on the acceptable number of immatures on leaf surfaces. In each PMU, plants are randomly selected and six leaves are inspected per plant.

The minimum number of plants inspected is 14 for the low threshold (0.1 nymph/sample unit), 10 for the moderate threshold (0.6 nymph/sample unit), and 6 for the high threshold (3.0 nymphs/sample unit). The cumulative number of each life stage of whitefly is recorded. If the cumulative number of nymphs in the PMU equals the maximum boundary (for the selected threshold), sampling is stopped and control measures are necessary that week for that PMU. If the minimum number of plants are inspected in the PMU and the cumulative number of whitefly nymphs is below the minimum boundary for the threshold, inspection is stopped, control actions are not necessary and that PMU is not inspected again until the following week.

Sequential sampling reduces inspection time and the cost of inspection, yet provides a high level of assurance that whiteflies are being controlled. The effectiveness of the sequential sampling plan was verified with the cooperation of commercial growers. Growers who used sequential sampling achieved their target control levels and reduced their insect scouting costs by 40 percent.

For a copy of the sequential poinsettia inspection plan contact Media Services Distribution Centre, Cornell University, 7 Business and Technology Park, Ithaca, NY 14850 (607-255-2080). The document title is "New York State IPM Program, publication No. 403," 1993, Cornell Cooperative Extension; the cost is $10.

Biological Control

Many commercial growers raise multiple crops, use continuous crop cycles, and need to control many pests simultaneously. In contrast, poinsettias are often the only crop in the greenhouse, have a single growing cycle, and are affected by only one pest, the whitefly. This makes poinsettias an excellent candidate for the use of biological control methods, which should be used as part of an overall IPM program.

Table 2. Action thresholds for whiteflies on poinsettia.

# plants sampled	Low threshold — A		Moderate threshold — B		High threshold — C	
	Upper limit	Lower limit	Upper limit	Lower limit	Upper limit	Lower limit
	—D	—E	—D	—E	—D	—E
2	0	-	3	-	15	-
4	1	-	5	-	25	-
6	1	-	7	-	34	2
8	1	-	9	-	42	6
10	2	-	11	1	50	10
12	2	-	12	2	58	14
14	2	1	14	3	66	18
16	3	1	16	4	74	22
18	3	1	17	4	81	27
20	3	1	19	5	89	31

22	3	1	20	6	96	36
24	4	1	22	7	104	40
26	4	1	23	8	111	45
28	4	1	25	9	118	50
30	4	2	27	9	125	55
35	5	2	30	12	143	67
40	6	2	34	14	161	79
45	6	3	38	16	179	91
50	7	3	41	19	196	104

Key

A Low threshold=0.1 nymphs/sample unit

B Moderate threshold=0.6 nymphs/sample unit

C High threshold=3.0 nymphs/sample unit

D Classify sample as "above threshold" if cumulative counts are equal to the upper limit

E Classify sample as "below threshold" if cumulative counts are less than the lower limit

From: Sanderson et al. 1994. Used with permission.

Example 1: Low Threshold. The number of nymphs are less than lower limit, therefore no action required this week. IPM scout randomly samples a pest management unit (PMU); the sampling plan calls out that at least 14 plants will need to be inspected before reaching the conclusion that no action is required. No nymphs were found after the completion of the required sample. Since zero is less than the lower limit of one, no action required this week.

Example 2: Low Threshold. The number of nymphs reach upper limit, management action required this week. IPM scout sets out to sample the PMU. If one nymph is found in any of the first 8 plants, the upper limit has been reached and a management decision is required. Similarly, if two nymphs are found on the 10th through 14th plant, a management decision is required. The scout found no nymphs until the 6th plant when 1 nymph was found, which meant that the upper limit was reached. A management decision is required this week for this PMU.

Example 3: Low Threshold. After initial target sample, the number of nymphs found is between the upper and lower limits. The sample is expanded until the number of nymphs is below the lower limit or equal to the upper limit. IPM scout randomly samples; no nymphs were found until the 14th plant on which a single nymph was found. The scout must continue sampling until a total of 30 plants have been inspected with no additional nymphs found. A single nymph in 30 plants is less than the lower limit of 2 nymphs, thus, no action required this week in this

PMU. Note that if the number of nymphs reaches the upper limit for any of the steps in the expanded sample, a management decision is required.

Entomopathogenic Fungi

Entomopathogenic fungi, also known as mycopesticides or mycopathogens, are fungi that prey on insects. Entomopathogenic fungi are a useful component of any IPM program because they are relatively host specific, inexpensive to produce, able to function in a wide range of greenhouse environments, and safe to humans. One type of entomopathogenic fungus, *Beauveria bassiana*, is very effective when whitefly populations are low. Three to five sprays typically eliminate whiteflies in the greenhouse.

Current research shows that three other entomopathogenic fungi, *Paecilomyces fumosoroseus*, *Metarrhyzium lecani*, and *Verticillium lecani* are effective at controlling whiteflies. These organisms, however, are either not available commercially or are not labelled for use in greenhouses.

Entomopathogenic fungi, however, do not offer stand-alone pest-control capabilities and are best used in conjunction with a program of conventional insecticides or insect growth regulators.

Natural Enemies

Research continues into controlling whiteflies in poinsettia production greenhouses using natural enemies of the whitefly. *Encarsia formosa*, a parasitoid, is effective against GWF but not SLWF. A predator, *Delphastus pussillus*, combined with the parasitoid, *Encarsia luteola*, was effective in trials against SLWF in commercial greenhouses, but controlling white flies with these organisms costs five times more than conventional pesticides. The parasitic wasp, Eretmocrus eremicus, when used in conjunction with insect growth regulators (IGRs), was effective against SLWF in experimental and commercial greenhouses. The advantage of using parasitoids in combination with IGRs is that fewer IGR applications are necessary, which reduces the probability that whiteflies will develop resistance to the IGRs.

Chemical Control

Imidacloprid, a member of a new class of synthetic insecticides called chloronicotinyls, has proven extremely effective against whiteflies in poinsettia crops. Marathon® 1G, which is labelled for ornamental and greenhouse crops, is a granular formulation of imidacloprid.

Imidacloprid is systemic, has a low mammalian toxicity, and is also effective against aphids. However, because imidacloprid is so effective and so long lasting, one application per crop can induce the whiteflies to develop resistance. To prevent imidacloprid resistance from developing, use other pesticides in alternate years in different areas of your greenhouse. Pesticides labelled for greenhouse use against whiteflies are shown in Table 3. Please note that pesticide classes are included in the table to help you plan pesticide rotations.

Table 3. Pesticides labelled for whiteflies on poinsettias.

Class	*Common Name*	*	*Trade Name*	*Effective On*
B	* azadirachtin (neem)		Azatin, Neemazad	Nm, Pp
B	* pyrethrum	1	PT 1100, X-clude	Nm, Ad
C	bendiocarb	4	Dycarb, Turcam	Nm, Ad
C	oxamyl		Oxamyl	Nm, Ad
C	fenoxycarb		Precision, Preclude	Nm, Pp
CH	endosulfan	2	Thiodan	Ad
CN	imidacloprid	2	Marathon	All Stages
H	* hort oils	3	Sun Spray, Ultra	Eg, Nm, Pp
IGR	diflubenzuron		Adept	Nm, Pp
IGR	kinoprene	1	Enstar II	All Stages
IGR	pyriproxyfen	1,2	Distance	Eg, Nm
Mb	Beauveria bassiana	6	Naturalis, Botanigard	All Stages
OP	chlorpyrifos		Duraguard	Nm, Ad
OP	acephate	1	Orthene, Pinpoint	Nm, Ad
OP	dichlorvos	4	DDVP	Nm, Ad
Soap	insecticidal soap	5	M-Pede, Safer	Nm, Pp, Ad
P	bifenthrin		Talstar, Attain	Nm, Ad
P	cyfluthrin		Decathlon	Nm, Ad
P	fluvalinate		Mavrik	Nm, Ad
P	fenpropathrin	1	Tame	Nm, Ad
P	lambda-cyhalothrin	2	Topcide	Nm, Ad
P	permethrin	2	Astr	*Nm, Ad
P	resmethrin		Resmethrin	Nm, Ad
Py	pyridaben	7	Sanmite	Nm, Ad

Key

1 Do not apply to bracts

2 Limited amount per crop

3 Some tank mix restrictions

4 Foliage and blooms must be dry

5 Only 3 consecutive applications

6 Incompatible with fungicides

7 Rotate 2 other products between

BO=Botanical, C=Carbamate, CH=Chlorinated Hydrocarbons, CN=Chloronicotinyl, HO=Horticultural Oils, IGR=Insect Growth Regulator, Mb=Microbial, OP=Organophosphates, P=Pyrethroids, Py=Pyridazinone

Eg=Eggs, Nm=Nymph, Pp=Pupa, Ad=Adult

Mention of a pesticide does not constitute an endorsement of any product and any omission from this list is unintentional. The pesticide label is the ultimate authority for pesticide use.

Entomology and Pest Management

Entomology is the scientific study of insects. Because they are closely related, we will include spiders, mites and *Arthropod* pests. Gardeners who understand pest life cycles and behaviour are better able to determine when control will be most effective. You probably have about 1,000 species of insects living in your yard. They are all part of nature's complex Ecosystems and Food Chains. Of the nearly 100,000 insect species in the USA, only about 600 species are considered pests. Since few insects are actual pests, Master Gardeners must learn which are pests, which are Beneficials, and which do not affect gardening. Insects play an integral part in our lives. Many gardeners are aware that beneficial insects are necessary to pollinate many fruit and vegetable crops. Other beneficial insects are the source of honey, beeswax, silk and shellac. Insects also help break down dead plant tissue. Some insects capture and eat other insects and are called predators. Other insects have larvae which attack pests by living inside their bodies and these are called parasitoids.

Pest insects and mites may feed on our plants and our homes, or they may carry diseases that infect plants. A few also carry diseases that infect animals, including people. Master Gardeners use the term "pest management" rather than "pest control" or "pest eradication." We now realize that we cannot eradicate or completely rid our gardens and homes of pests. We can only try to keep pest numbers low to minimize damage.

Pest management is a complex job that involves much more than just spraying with pesticides. Master Gardeners must learn how to use other pest management techniques. You will learn how to use beneficial insects (Biological Controls). You will also learn how to make homes and gardens less attractive to pests (Cultural Controls).

Arthropods

Insects are a class or subgroup of a larger group called of phylum, known as arthropods (ARE-throw-pods). This word means "jointed foot." The arthropod group includes insects, spiders, crayfish, millipedes and similar animals. The various groups within the arthropods are distinguished from each other by the arrangement of their body parts.

Insects

Insects are land animals with 3 pairs of jointed legs. The insect group includes many subgroups: termites; grasshoppers, crickets and roaches; earwigs; lice; true bugs; beetles and grubs; butterflies, moths and caterpillars; fleas; flies, gnats and maggots; and bees and wasps.

The word "insect" comes from a Latin word which means "to cut into." This describes how the bodies of insects are divided into distinct regions: head, thorax and abdomen.

People often refer to insects as bugs. This is incorrect. Bugs are a subgroup of insects, just like the beetles, flies, bees and wasps.

Body Parts

There are three characteristics found among all insects:

3 body regions: head, thorax (THOR-ax) and abdomen

1 pair of antennae (an-TEN-e)

3 pairs of legs attached to the thorax

A typical insect is shown in the following illustration.

None of the arthropods, including insects, have a skeleton of bones. Instead, they have a tough outer shell called an exoskeleton. The exoskeleton provides support and protection. In order to grow, arthropods must shed the exoskeleton or molt.

The Head

The insect head contains sensory organs, such as the antennae, eyes and mouth. The antennae on an insect are usually easy to see. One pair is always present on the adult form (except scale insects). Though you may call insect antennae "feelers," they are used by the insect to smell. The antennae can be very helpful in identification. They vary by species in number of segments and shape.

Insect eyes are very different from human eyes. Insect eyes do not form an image like our eyes do. Insect eyes are very good for detecting motion which helps the insect escape danger.

Insects have two basic types of mouthparts: chewing and sucking. There are variations of these two basic types. Thrips have RASPING-sucking mouths. Honeybees, wasps and bumblebees have chewing-lapping mouths. Aphids, scales and mealy bugs, have piercing-sucking mouths. Many of us know firsthand about the blood-sucking mouths of fleas, mosquitoes and lice. The different mouthpart types cause characteristic types of damage which can be used as clues for identification of the pest. Just as you can identify mammals by their footprints, you can determine the kind of insect causing damage to a plant by the feeding print left by the mouthparts.

The Thorax

The thorax is the middle body region. The legs and wings are attached to this region. Number of wings, the VENATION and the presence of wing covers sometimes will be used in identifying insects. Most adult insects have 2 pairs of wings, though some have lost 1 pair (flies) or even both pairs (fleas). Only adult insects have wings.

The 3 pairs of legs on insects are always present on adults. They may or may not be present on immature stages.

The Abdomen

The last body region is the abdomen. It consists of 11 or fewer segments. The number of segments is often helpful in identification.

It is a good idea for the gardener to purchase and use a small magnifying or hand lens. Use it to look for insect eggs and to examine plant damage and very small insects. A hand lens helps with critical distinctions among many common insect pests.

Life Cycles

Most gardeners can recognize GRUBS and CATERPILLARS. However, they may not be aware that they are the immature stages of common insects. Insects are very different from other animals. They often progress through very different forms called stages, as they develop into adults.

Understanding insect life cycles is important for management of pests. Many insects are easy to manage during one or two of their stages. Pesticides may not be effective against insect eggs or adults, while the soft immature insect may be easily killed. It, therefore, becomes important to recognize these pests in immature stages of their life cycle.

Metamorphosis

Metamorphosis (met-a-MORE-feh-sis) is the unique developmental process in insects. Metamorphosis is a combination of two Greek words: “meta” which means change, and “morpho” which means form. Complete or complex metamorphosis has the following stages: egg, larva, pupa and adult. We are probably most familiar with this process in butterflies and moths.

Complete Metamorphosis

Egg: Insects lay fertilized eggs singly or in clusters. At this stage, vulnerability to chemical control is minimal. The egg has a protective covering. These clusters of eggs often are enclosed in a protective material. However, egg clusters can be removed and crushed. Eggs may be laid in protected, hidden locations to avoid exposure to weather and predators.

Larva: When an egg hatches, the larva emerges. Larvae are the feeding, growing, nonreproductive stage of development. During this stage most insects go through a process called molting. Instead of gradually growing larger, the larva eats until its skin cannot expand any further. The old skin splits and is shed. The stage between each molting is called an instar.

Following each molting stage or instar, feeding increases. In the garden, this can be noticed by increased plant damage. It is during the larval stage that most insects cause plant damage. Caterpillars, grubs and maggots are larval stages of moths and butterflies, beetles and flies, respectively.

Control is usually easiest during the larval stages. The larva is feeding and is more likely to ingest pesticides. During molting, the body is exposed to contact pesticides. Infection by natural diseases and attacks by predators or parasitoids are also more likely.

Pupa: (PEW-pah) is a non-feeding, transforming stage. This stage may form a kind of covering, such as the cocoon or chrysalis made by the butterfly. During this stage the insect transforms into the adult stage, often an entirely different form from the larva. At this stage, the only possible control is mechanical, such as crushing.

Adult: Adults are the reproductive stage of insects. They may or may not feed. Some live for a very brief period, even one day or less. Control is possible during this stage of development, but usually more difficult. It is critical to control pests before they reproduce a new generation.

Incomplete Metamorphosis

Many insects do not develop through a complete metamorphosis. Instead they become adults through what is called incomplete or simple metamorphosis. Simple metamorphosis involves a gradual change in size, but not form.

The intermediate stage is called the nymph. Nymphal stages look like the adult, but they may have different colour markings and the wings are merely pads visible on the thorax.

When the last nymphal instar molts, the newly emerged adult expands the wings. The following illustration shows incomplete metamorphosis.

How Insects Survive Winter

Many insects overwinter in a dormant state. They can be in any stage of development: egg, larva, pupa or adult. In the garden, insects spend the winter in soil or dead plant material. Rough tilling of the soil can expose the insect to fluctuating temperatures that may kill it. Removal and disposal of plant debris also may eliminate overwintering insect stages.

Social Insects

Certain insects have developed a unique system of living. They live in a social colony. There are different forms within the colony that handle different duties. Each group performs separate duties that ensure survival of the colony. Often, a colony consists of winged reproductives, called queen(s) and drone(s), workers with or without wings, and young that often are completely helpless.

In some social insect groups there is a king, or drone, and soldiers which protect the group. Eggs and immatures are cared for and protected by other colony members. The queen usually has the sole job of laying eggs and establishing a new colony, if necessary. Examples of social insects include ants, termites, wasps, and bees.

Insect Pests of Houses, Pets, and Humans

Household Pests

Many pests found in the house can be controlled by excluding them from the house with screen and filling all cracks and crevices. Other household pests can be controlled by sanitation.

Pantry Pests

Pantry pests are a group of insects that feed on dried vegetable and animal matter. This can include flour, pasta, spices, cereals, powdered milk, dry cat and dog food and birdseed. Common pantry pests include the following:

- Meal moths
- Flour moths
- Grain moths
- Saw-toothed beetles
- Merchant grain
- beetles
- Confused beetles
- Red flour beetles.

These insects are often first noticed crawling or flying around the kitchen or in areas where food is stored. Careful food purchase, storage and sanitation are critical to control. These insects do not transmit disease, but can make many foods unusable. Fact Sheets 2086, 2087 and 2089, References Ento.7, Ento.8 and Ento.9, give information on identification and controls for each insect.

Vinegar Flies

Vinegar flies or fruit flies, become a nuisance in homes when they fly around overripe fruit and vegetables. They may also be attracted to food that contains yeast, such as vinegar, beer, wine and cider. They may congregate around any area where there is garbage or food-contaminated water. Some can carry disease from human and animal excrement to foods.

To control vinegar flies, concentrate on eliminating sites where larvae feed and adult flies breed. Sanitation is critical to control. Fact Sheet 2109 gives details on preferred foods, sites and sanitation measures for control. It also provides a homemade trap and suggests appropriate insecticides.

Cockroaches

Roaches are undesirable insects because they can foul and contaminate food, carry diseases, damage paper and cloth products, and produce unpleasant odors. They can be very persistent pests once populations become established in a dwelling. This happens because roaches can live long periods without food.

The most common roach in urban areas is the German cockroach. German roaches live throughout dwellings, but are often found in kitchens and bathrooms. Because this roach produces more eggs and generations per year than other roaches, it only takes a few roaches to develop quickly into a major infestation. If German roaches are seen by day, it means that a large population exists and all hiding places are taken. Fact Sheet 2099 provides information on identification and control of this roach, Reference Ento.10.

Another type of roach, the oriental cockroach, prefers damp areas and lives both indoors and outside. In dwellings, these roaches can be found in locations where there is dampness or moisture. They stay mostly in the basement and on the first floor. In warm weather they live outside and may migrate indoors when it becomes cool. Fact Sheet 2097, Reference Ento.11, gives details on identification and control of oriental cockroaches.

The brown-banded cockroach prefers to hide in warm, elevated areas near the ceiling within dwellings. However, it will develop and live throughout a building. This makes control difficult. Fact Sheet 2098 gives details on where to look for infestations of this roach and control, Reference Ento.12.

The largest roach, the American roach, is found throughout the house. It is often called a water bug.

Sanitation is critical to the control of all household insect pests that feed on foodstuffs.

- Clean up all spilled food.
- Store food in sealable glass, metal or plastic containers.
- Rinse all food packaging before disposal.
- Use insect-proof containers for garbage both indoors and outdoors.
- Promptly treat any infestation.
- Dispose of vacuum cleaner bags after cleaning infested areas.
- Avoid foods in broken packaging.

Use the Fact Sheets for control recommendations. In cases of severe infestations, a reputable pest control firm may be advisable. Fact Sheet 2091, Reference Ento.13, gives guidelines for choosing a pest control company.

Wood Destroyers

Several insects use firewood and structural wood for their homes and as a food source. They include flatheaded borers; old house borers; bark, powderpost, and darkling beetles; carpenter ants; termites and carpenter bees. Refer to the Fact Sheets on these pests when problems arise. Fact Sheet 2065, Reference Ento.14, "Firewood Insects and Their Relatives," provides guidelines for controlling this seasonal problem.

Infestations of carpenter ants and termites cause alarm and panic among homeowners. Do not feel pressured if these pests are discovered. These insects work slowly. A house cannot be ruined overnight. There are two important aspects in the control of wood destroying pests:

- treatment of the current pest population;
- changes that will bring about long-term control, including structural changes.

Carpenter Ants

Carpenter ants are efficient wood-destroying insects. They are generally large compared to other ants.

Carpenter ants do not eat wood. They excavate it for their nest. They eat fruits, juices, meats, fats, dead insects and Honeydew created by other insects.

Carpenter ants can seriously damage wood structures. They normally nest in decaying wood that is easy to invade. Carpenter ants can nest in or just outside a house and invade rooms in search of food. Carpenter ants infest undisturbed piles of damp or rotting firewood stacked outside on the ground. They also nest in the rotting wood of trees. Do not store wood in damp basements or on the ground against the house.

Termites

Subterranean termites are found in Ohio. These termites are social insects that live in colonies in the soil. Termites cause structural damage as workers feed on wood. They chew wood and convert it to food for other colony members. Subterranean termites usually limit their attack to wood that is near soil. However, they may build tubes running from the soil to wood. These tubes are usually in protected basements, crawl spaces or in cracks in the foundation. Moisture is essential for their survival.

To avoid termite infestation, do not bury wood near buildings. Remove old tree stumps near buildings. Provide drainage away from structures. Do not allow wooden portions of structures to contact soil. Ventilate crawl spaces under structures. For typical signs of termite infestation, termite identification, and recommended control measures, refer to Fact Sheet 2092, Reference Ento.17.

Pest Management

There have been and are many approaches to managing pests in crops, gardens, yards and homes. These range from applying no control measures to applying pesticides on a regular schedule. Some people will not use pesticides; others want to get rid of the pest by any means. As a Master Gardener, you must learn all the methods and materials which can be used to manage pests.

When we try to understand all the methods of managing pests and select the best methods for a particular situation, we are using Integrated Pest Management (IPM).

Integrated Pest Management (IPM)

Integrated Pest Management or IPM provides us with a systematic approach to pest control. It begins with using the best gardening practices. Following the correct selection and culture of plants reduces the potential for pests and diseases.

When problems do arise, the next step in an IPM program is determining when a pest is causing enough damage to warrant control. It also helps the gardener determine which of the many possible controls is most effective and appropriate in a particular situation. IPM can be applied to all pest groups: insects, mites, weeds, plant diseases and animals.

Correct Cultural Practices

Cultural practices are ways the gardener has of changing environmental factors that affect the plants and their pest populations. Cultural practices include irrigation, crop rotation, garden sanitation, soil aeration, mulching, tilling and use of disease- and pest-resistant plants. It is critical that gardeners know the cultural or growing requirements of each plant.

Providing the correct growing conditions results in a vigorous plant that is less likely to be attacked by pests and diseases. Vigorous plants also can tolerate some damage.

The question of when an organism becomes a pest is central to IPM. Deciding how much damage is acceptable with a particular plant or crop is called "establishing a threshold."

Establish Thresholds

Several factors influence how each gardener will arrive at a pest threshold. These factors include the following:

- amount of damage that can be tolerated
- pest population size required to cause significant damage
- the plants' stage of development
- the vigour of the threatened plant(s)
- the life cycle and habits of the pest.

The amount of damage that can be tolerated varies with the crop and personal preferences. A few holes on a collard leaf does not require control. However, if most of the leaf has been eaten, the crop is destroyed.

Less than 12 white grubs in a square foot of turfgrass can be tolerated as long as the lawn is growing vigorously. However, if the density of grubs rises above 12 per square foot, damage will be significant.

Colorado potato beetles will devastate a potato crop early in the season. Late in the season the beetles can ea t all of the foliage and there is little damage to the crop.

Peach tree borers are extremely difficult to control once they bore into the trunk of the tree. Knowing that the adults lay eggs in mid-July to mid-August and that the young larvae are susceptible during this period, permits the gardener to effectively apply controls.

Establishing thresholds in the garden involves observing the plant and its total environment.

There is also the question: Does the cure cause more damage than the pest? There is a risk involved in using the more toxic pesticides. This risk must be weighed against the potential benefit.

Inspect and Monitor

You must monitor the garden to determine when action is necessary. Thoroughly inspect the plant: the underside of leaves, buds, new growth, even the roots in extreme cases.

Regular inspection allows the gardener to identify a problem before major damage occurs to the plant. Controls will be more effective if the problem is caught early. Infrequent inspection results in extensive damage. Once extensive damage occurs, it is often too late to do anything.

Inspect the plant's entire environment for clues to the problem. Too much or too little moisture, wind, humidity and light may cause stress that results in susceptibility to damage from pests and diseases. Observing and keeping records of weather conditions provide clues to growth patterns and problems.

Certain insect pests can be monitored by using traps. Coloured, sticky traps, PHEROMONE traps, pitfall traps and light traps are used with some pests. Whiteflies and aphids are attracted to bright yellow, sticky cards and are trapped on the sticky surface.

Identify the Cause of the Problem

The problem may be caused by insects or related pests, diseases or cultural practices. Often, the problem is caused by a combination of these factors. Identify the pest or disease and the conditions that allowed that organism to become a problem.

Reference materials will guide you to common problems for a particular plant. Most guides are organized by plant. They list the insects and diseases that attack each plant. The ORTHO PROBLEM SOLVER is an excellent resource for identification. A questioning attitude and close inspection are required to make an accurate diagnosis. A magnifying or hand lens is helpful.

Closely examine the specimen for feeding or disease patterns. If insects are involved, try to collect the pest. Identification of insects and diseases is available through local Extension offices and the Plant and Pest Diagnostic Clinic at The Ohio State University.

Apply Controls

To prescribe a control, you must understand the pest organism, its life cycle and the type of damage it causes. Consider all control options. Select the most effective and appropriate option or combination of options. Apply controls at the right time. Insects are more vulnerable at certain stages of their life cycle. Eggs may be controlled with oil sprays. Larval stages are particularly vulnerable. It is critical to control insects before they reproduce and increase their population. When an insect is protected by the pupal case, control is nearly impossible.

Reapplication may be necessary for many pests to kill the young as they emerge from eggs. Know how long the pesticide remains effective on the plant. This is also important in food crops. You must observe the prescribed time lapse between the last application and harvest.

You may want to apply a pesticide and then enclose the plants in netting to keep further infestation from occurring. Or, you may decide to put up netting and then release predators.

Diseases are more difficult to control. Most of the fungicides prevent infection or at best stop an infection from spreading.

Apply controls as instructed. Netting is not effective if there are unprotected areas. Pesticides must be applied to both the top and underside of leaves. Surfactants or spreader/stickers help the pesticide adhere to and spread over the foliage surface. Application rates are critical. The proper dilution is required to kill the pest. If too concentrated, plant damage may occur.

Submitting Insects For Identification

When an insect pest can't be identified using a fact sheet, a field guide or other resources, it should be submitted to the Plant and Pest Diagnostic Clinic. Fact Sheet 2121 provides tips on specimen preparation. Consult Extension's office staff for the current fee. Payment must accompany all specimens.

Orders of Insects and Their Feeding Parts

The following is a condensed list of insects by order. These orders include insects that are important to home gardeners. This list indicates how each group feeds on plants and the type of metamorphosis.

Coleoptera: beetles and weevils

Mouthparts: chewing

Metamorphosis: complete

Dermaptera: earwigs

Mouthparts: chewing

Metamorphosis: incomplete

Diptera: flies, gnats, midges, mosquitoes

Mouthparts: sponging and piercing/sucking

Metamorphosis: complete

Hemiptera: stinkbug, plantbug, squash bug, boxelder bug

Mouthparts: piercing/sucking

Metamorphosis: incomplete

Homoptera: scale, mealybugs, whiteflies, aphids, cicadas, leafhoppers

Mouthparts: piercing/sucking

Metamorphosis: incomplete

Hymenoptera: bees, ants, wasps, sawflies, horntails

Mouthparts: chewing

Metamorphosis: complete

Isoptera: termites

Mouthparts: chewing

Metamorphosis: incomplete

Lepidoptera: butterflies and moths

Mouthparts: larvae-chewing

Adults-sucking

Metamorphosis: complete

Neuroptera: lacewings, antlions, alderflies, dobsonflies

Mouthparts: chewing

Metamorphosis: complete

Orthoptera: crickets, praying mantids, grasshoppers

Mouthparts: chewing

Metamorphosis: incomplete

Siphonaptera: fleas

Mouthparts: larvae-chewing

Adults-sucking

Metamorphosis: complete

Mouthparts: rasping/sucking

Metamorphosis: complete and incomplete.

6

Entomology and Ganglion Cells

General Morphology

Ganglion cells are the final output neurons of the vertebrate retina. The ganglion cell collects the electrical messages concerning the visual signal from the two layers of nerve cells preceding it in the retinal wiring scheme. A great deal of preprocessing has been accomplished by the neurons of the vertical pathways (photoreceptor to bipolar to ganglion cell chain), and by the lateral pathways (photoreceptor to horizontal cell to bipolar to amacrine to ganglion cell chain) before presentation to the ganglion cell and so it represents the ultimate signaller to the brain of retinal information.

Ganglion cells are larger on average than most preceding retinal interneurons and have large diameter axons capable of passing the electrical signal, in the form of transient spike trains, to the retinal recipient areas of the brain many millimetres or centimeters distant from the retina. The optic nerve collects all the axons of the ganglion cells and this bundle of more than a million fibres (in man at least) then passes information to the next relay station in the brain for sorting and integrating into further information processing channels.

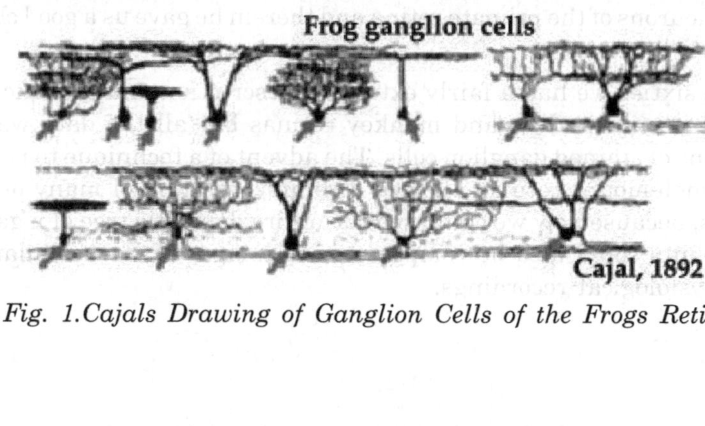

Fig. 1.Cajals Drawing of Ganglion Cells of the Frogs Retina

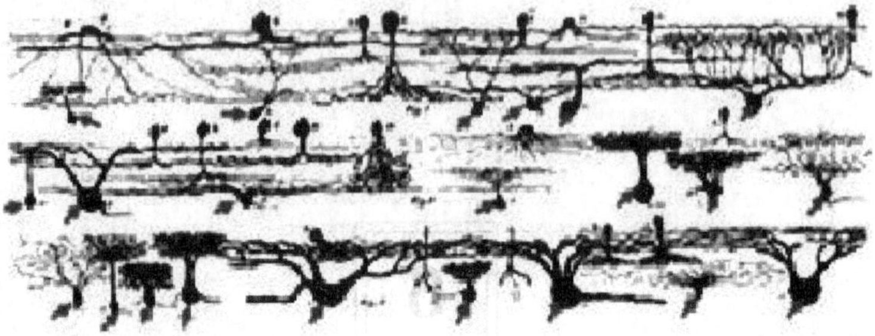

Fig. 2. Cajals Drawing of Ganglion Cells of Ox and Dog Retinas

Cajal in his monumental work on Golgi staining of the vertebrate retina was able to classify many different varieties of ganglion cell based on form (dendritic morphology), extent (cell body and dendritic tree size), and number of sublayers in which they arborize (stratification levels in the inner plexiform layer).

He considered the retina to be remarkably uniform across all vertebrates differing only in respect to rod and cone specializations for the visual sense of the animal.

Looking at Cajals (1892) drawings of dog ganglion cells as compared frog ganglion cell for example the former seem simpler in form than the latter cells but we actually now know that there is a common evolutionary path taken by different ganglion cell types so that different morphological and functional classes are similar throughout the species.

For example the large ganglion cells, with open radiate branching patterns, process fast, transient impulse trains and in all vertebrate retinas are concerned with motion detection and alerting the animal to threatening, moving visual imagery.

While small bushy ganglion cell types are concerned with processing small stationary, fine detail in tonically activated messages in all species.

In the forties, Polyak (1941) produced a phenomenal description of the Golgi-impregnated neurons of the primate retina and therein he gave us a good classification of ganglion cell types.

So by the sixties we had a fairly extensive description and classification of the ganglion cells in mammalian and monkey retinas but all the data was based on vertical sections of stained ganglion cells. The advent of a technique to perform Golgi staining on wholemount retinas allowed a reinterpretation of many of the earlier classifications, because now we could see the entire dendritic tree of a ganglion cell. The wholemounts could also be compared with images of intracellularly injected cells after physiological recordings.

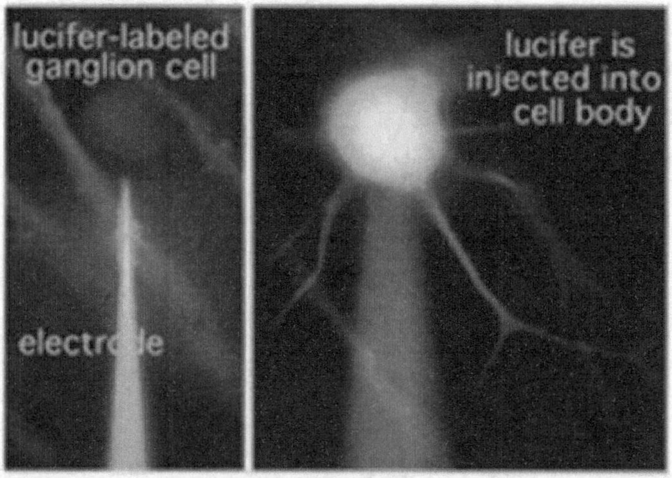

Fig. 3. Lucifer Filled Electrode Records and Stains a Ganglion Cell

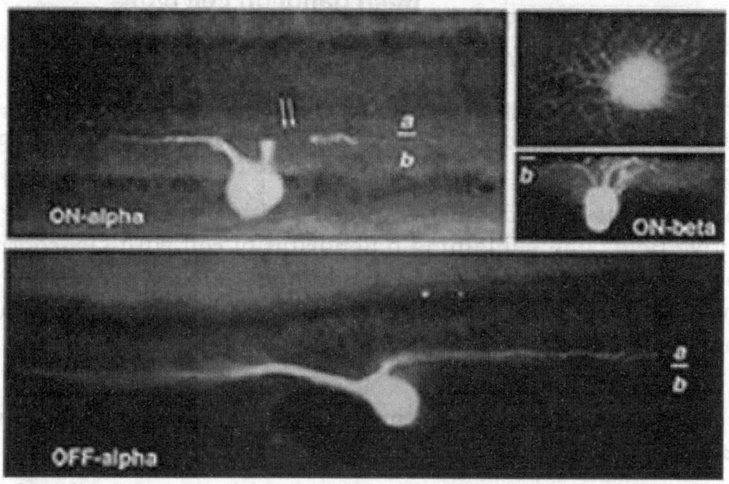

Fig. 4. Lucifer Injected and Intracellularly Recorded Cat Ganglion Cells

A particularly successful morphological classification scheme was proposed by Boycott and Wassle (1974). Three main classes of physiological response type were correlated with three morphological classes of ganglion cell in cat retina. Thus the alpha, beta, gamma and delta morphological ganglion cell types were considered to be the equivalents of the Y, X and W types of the physiology. By comparison with Cajals drawings, it was clear that alpha and beta types were already described and a further ten or more of Cajals (1892) types were lumped under the umbrella of gamma and delta groups.

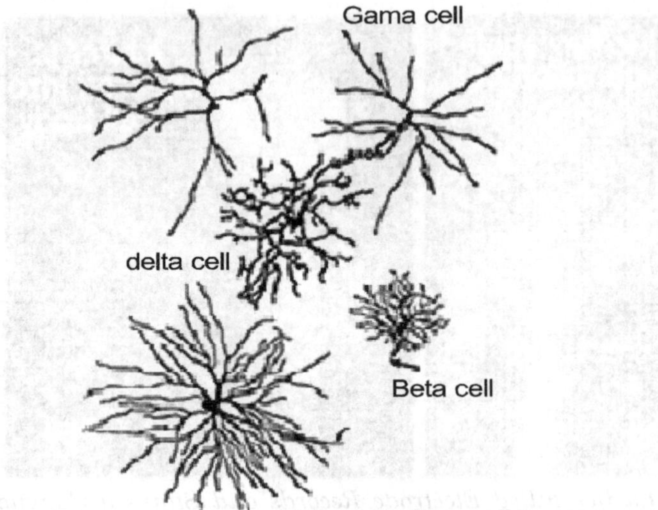

Fig. 5. Golgi stained cat ganglion cells

By 1978, we knew that alpha and beta cells of Boycott and Wassle (1974) could be subdivided into separate subtypes depending on whether they branch in sublamina a or sublamina b of the inner plexiform layer.

Sublamina a contained dendrites of cells with physiologically defined OFF-centre receptive fields and sublamina b dendrites of cell with ON-centre receptive fields. A new classification of Golgi stained ganglion cells could now relate both alpha and beta cells to the functionally important sublamina scheme and further, described many new types of cell including and going beyond the gamma and delta cells.

The 20 cell types over and above the alpha, beta and gamma cells were named from G4-G23 in order of cell body size and dendritic form.

Alpha and Beta Ganglion Cell Types of Mammalian Retinas

Alpha and beta ganglion cells of the cat retina are arranged in regular, superimposed bi-level mosaics across the whole retina. Both OFF-centre (branching in sublamina a) and ON-centre (branching in sublamina b) varieties of alpha and beta cells are arranged in a closely packed pattern with minimal overlap of their peripheral dendrites.

ON- and OFF varieties are always paired as shown in Wassle and coauthors elegant papers (1984), and between an ON/OFF pair there is sampling of almost the same retinal space (Fig. 7).

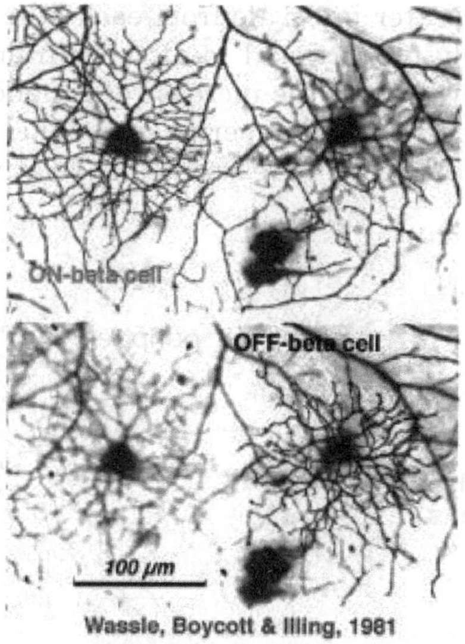

Fig. 6. On-and Off-Centre Beta Cells of Cat Retina Shown at Different Planes of Focus

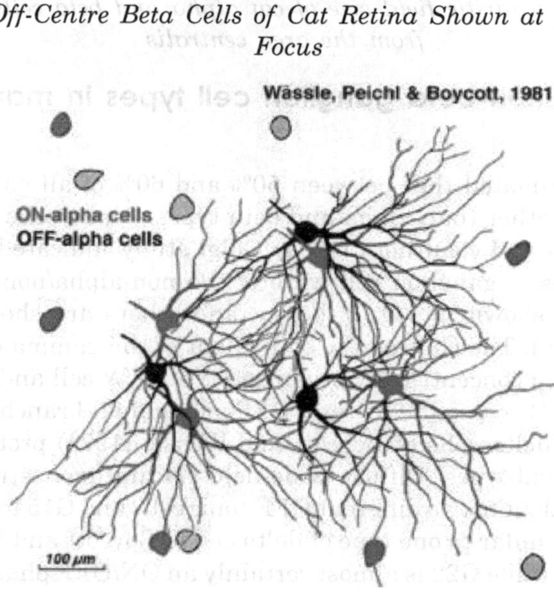

Fig. 7. HRP-Filled On-and Off-Centre Alpha Cell Mosaics in Cat Retina

Both cell types have their smallest dendritic tree sizes at the area centralis of the cat retina (equivalent to the fovea of primate retinas) and radiate out from this central point with gradually expanding circular dendritic trees. To maintain the low overlap ratio of their dendritic trees, the cells with larger fields are consequently spaced at greater intervals from each other, i.e. density of the cells/ mm2 of retina is inversely proportional to distance from the area centralis. This is shown in figure 8 (below) where the increase of ganglion cell size as shown with distance from the area centralis (a.c.): peripheral alpha and beta retinal ganglion cells are ten fold larger than their area centralis relatives.

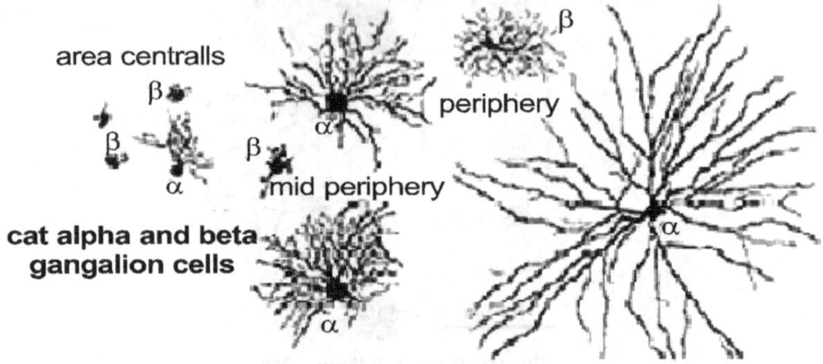

Fig. 8. Increase in dendritic field size of cat alpha and beta cells with eccentricity from the area centralis

Non-alpha and non-beta ganglion cell types in mammalian retinas

It has been estimated that between 50% and 60% of all ganglion cells in the cat retina are cells other than alpha and beta types. Alpha cells form 3% and beta cells 45-50 %. Kolb and coauthors (1981) Golgi study indicated that 21 different morphological types of ganglion cell were of the non-alpha/non beta cells classes (some of them are shown in Fig. 9, below, and others are shown in the human ganglion cell section). The G3 was the equivalent of the gamma cell of Boycott and Wassle. G3 may be a concentrically organized type of W cell and according to Kolb and coauthors (1981) comes in ON- and OFF-sublaminal branching subtypes. G19 cells equivalent to delta cells of Boycott and Wassle (1974) probably also come in the ON- and OFF subtypes. Diffuse wide-field G4 and monostratifying G18 cells are probably equivalent to two different OFF-tonic cells and G15 to an ON-directional cell. G19 which is similar to one type of delta cell of Boycott and Wassle (1974) may be an ON-tonic cell while G22 is almost certainly an ON/OFF phasic type. The newly described Q cell of Troy et al., (1995) may also be the physiological correlate of an ON-delta morphological type. Ganglion cells with the morphology of G21 and G23 are labelled by retrograde transport of tracers from brain visual centres: G21 from the accessory optic system. The speculation is that G21 is a directionally selective ganglion cell because cells of the accessory optic system are reported to be directionally selective. A ganglion cell type that could be a G23 type has been labelled from retrograde markers injected into the retinal recipient zone (RRZ) in the pulvinar and has been named the epsilon cell. However, the staining of epsilon cells has never been good enough to be sure which Golgi type it really is.

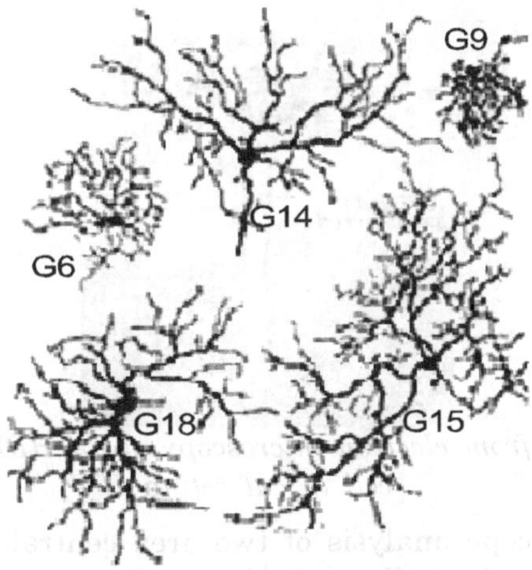

Fig. 9. Some of non-alpha, non-beta ganglion cells of cat retina

Other ganglion cell types that have been reported using tracer or other staining techniques than the Golgi method, include the medium-gamma cells, g1 and g2 cells , Neuropeptide Y immunoreactive (NPY-IR) cells, ganglion cells of the geniculate wing , and indoleamine containing ganglion cells likened to delta cells.

Convergence of Photoreceptors and Second Order Neurons to Cat Retinal Ganglion Cells

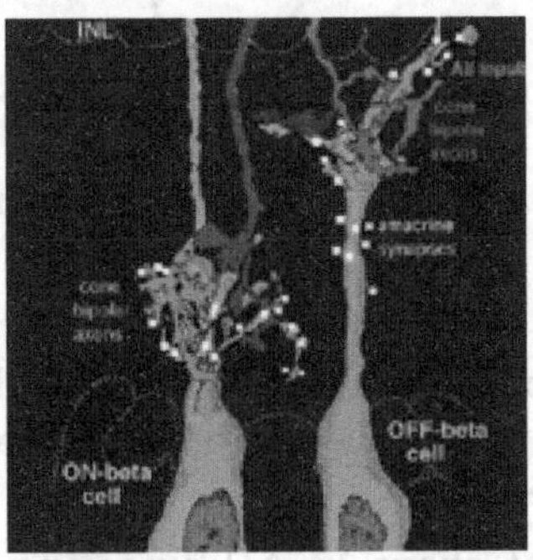

Fig. 10. Reconstructions from electron microscopy of ON- and OFF-centre beta cells of cat area centralis

The organization of the cat retina with a concentration of small-field beta cells in the area centralis allows the lowest convergence ratio of cone photoreceptors to ganglion cells. This is presumed to be the substrate for the high acuity pathways. The human (primate) retina culminates this trend with the development of the midget system and the photoreceptor to ganglion cell pathway to a one to one relationship of course.

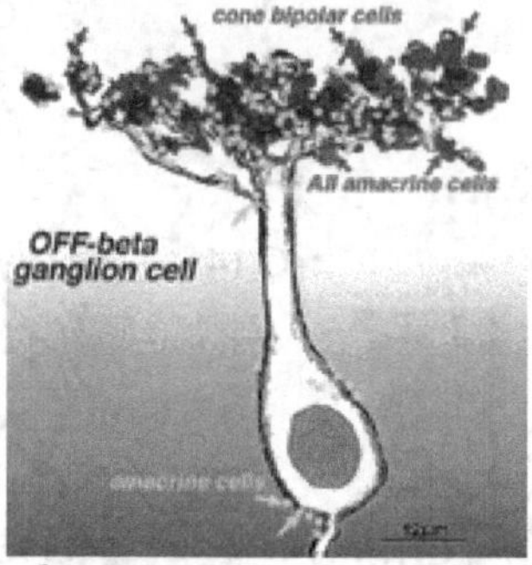

Fig. 11. Reconstruction from electron microscopy of an HRP-filled OFF-centre beta cell in cat retina

An electron microscope analysis of two area centralis beta cells in the cat, concluded that each ganglion cell received input from 3 or 4 cone bipolars which, in turn, received input from 4-8 cones.

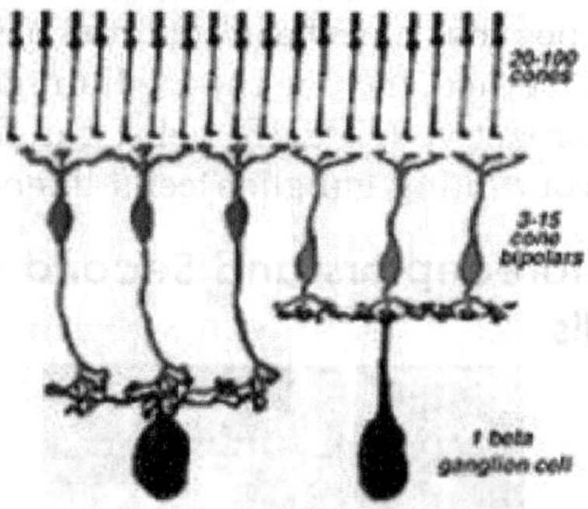

Fig. 12. Convergence of cones and bipolar cells upon ON- and OFF-centre beta cells

Another OFF-centre beta ganglion cell has been calculated to have fourteen definite bipolar cell axon terminals inputting the dendrites. Thus this type of ganglion cell would have convergence of approximately 100 cones through the 14 bipolar cells of two subtypes.

It is also calculated that about 14 AII amacrine cells, would also have input thereby converging 4200 rods through 280 rod bipolar cells to the 14 AII cells of input.

In addition, 28 to 41 other amacrine cells probably have synapses allowing even more convergence over a greater laterally spreading area than the ganglion cells dendritic tree diameter would have. In the same vein, we have information on convergence of bipolar cells and amacrine cells upon both ON-centre and OFF centre alpha ganglion cells.

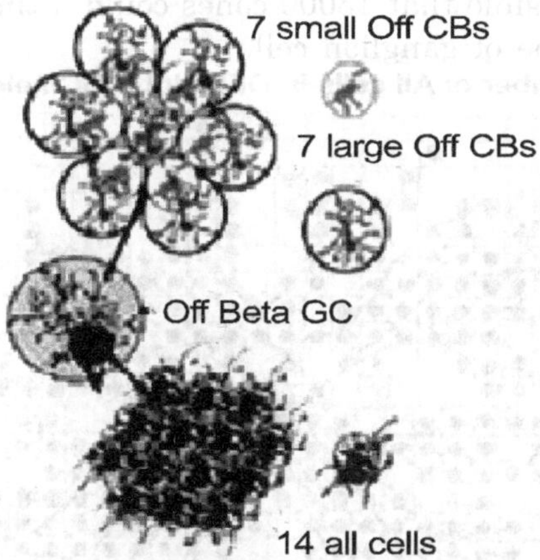

Fig. 13. Model of the two cone bipolar types and AII amacrine cell inputs to the dendritic tree of an OFF-beta cell

In the model of the OFF-centre alpha cell seen below, cone bipolar synapses, from only one type of OFF-centre cone bipolar (cb2 type, red dots), form 20% of the total synaptic input and the rest of the input is from various kinds of amacrine cells.

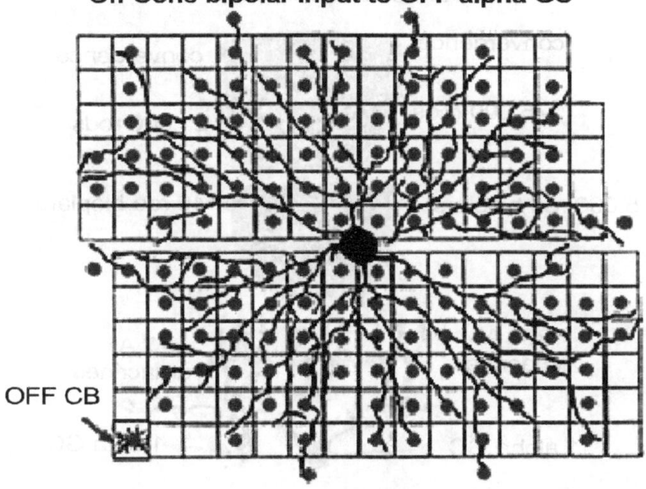

Fig. 14. Model of numbers of cone bipolar cells inputting an OFF-centre alpha cell

A minimum of 142 bipolar cells, 256 AII amacrine cells and 1011 other amacrine cells, together, provide between 6000 and 10000 synapses on the dendritic tree of this OFF-centre alpha ganglion cell. With a potential cone convergence of 5-15 cones per bipolar cell it is possible that 15000 cones could ultimately be involved in the capture area of this type of ganglion cell.

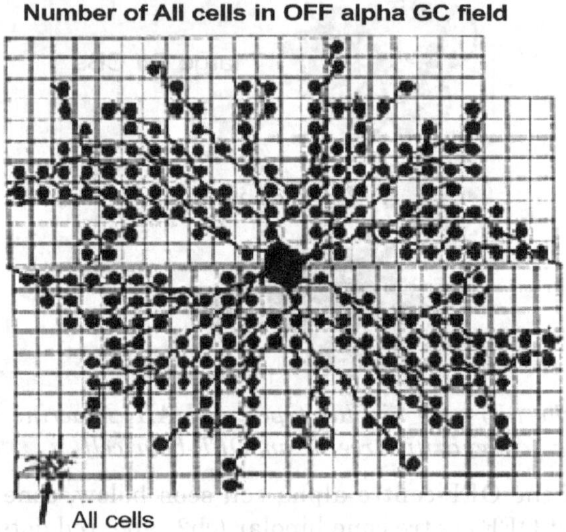

Fig. 15. Model of the number of AII amacrine cells inputting the dendritic tree of an OFF-alpha cell

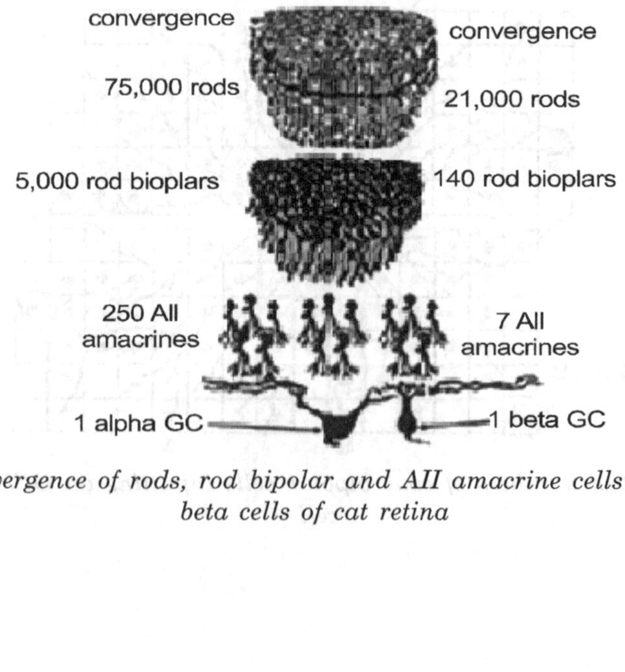

Fig. 16. Convergence of rods, rod bipolar and AII amacrine cells to alpha and beta cells of cat retina

In the case of the rod system, the convergence of preceding neurons to the ganglion cell is enormous i.e calculated by Sterling et al (1988) to be 75,000 rods through 5000 rod bipolar cells to 250 AII amacrine cells to a single alpha cell. The other amacrine cell input and their corresponding convergences is not capable of being assessed at the present time. Interestingly though, physiological measures of receptive field centre sizes of alpha cells are not much larger than their dendritic tree sizes despite enormous convergence and very wide lateral interactive potential.

Identity of Amacrine Cell Types Inputting Cat Alpha and Beta Ganglion Cells

As shown in the summary diagrams recent evidence suggests that several different morphological types of amacrine cells, as well as bipolar cell types, have synaptic input to the OFF-beta and OFF-alpha ganglion cells.

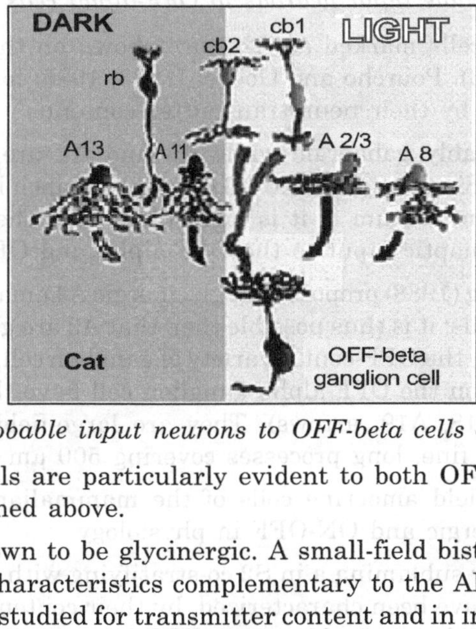

Fig. 17. Probable input neurons to OFF-beta cells of cat retina

AII amacrine cells are particularly evident to both OFF-centre ganglion cell types and as mentioned above.

AII cells are known to be glycinergic. A small-field bistratified amacrine cell with morphological characteristics complementary to the AII known as A8 (Kolb, et al., 1981) has been studied for transmitter content and in intracellular recordings and marking. A8 is considered a glycinergic cell. It synapses in sublamina a upon tufted dendrites and dendritic varicosities typical of the OFF-beta ganglion cell.

However, these A8 amacrine profiles have not been seen to input large diameter OFF-alpha ganglion cell dendrites (Fig. 18). A13 cells are considered GABAergic cells and known to be postsynaptic to rod and cone bipolars and make synapses with the cell body and primary dendrites in sublamina a and b of both ganglion cell types as well (Figs. 17 and 18, A13).

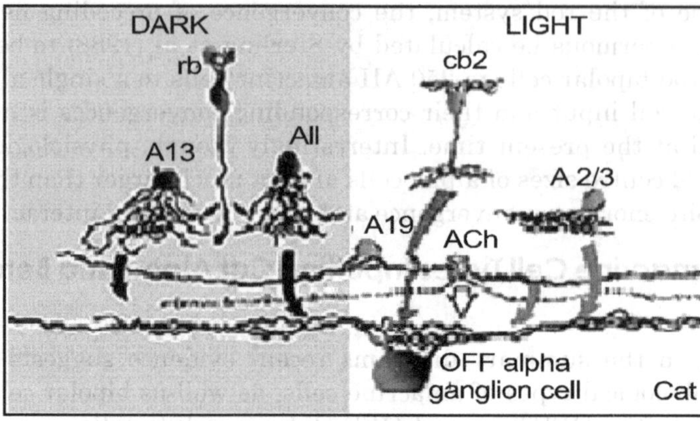

Fig. 18. Probable input neurons to OFF-alpha cells of cat retina

Small amacrine cells marked as A2/3 are shown on the summary diagrams (Figs. 17 and 18, A2/3). Pourcho and Goebel (1983; 1985) have clearly subdivided types A2, A3 and A4 by their neurotransmitter content.

Thus A2 is probably gabaergic while A3 and A4 are apparently medium-intensity glycinergic. Since only A2 and A3 cell types branch in strata 1 and 2 while A4 branches deeper in stratum 3, it is more likely that the former two cells are the candidates for synaptic input to the OFF-alpha and OFF-beta cells.

Freed and Sterling (1988) propose that glycinergic A4 amacrines are presynaptic to ON-centre alpha cells: it is thus possible then that A3 are glycinergic counterpart amacrines reserved for the OFF-centre variety of ganglion cell. Some of the amacrine profiles synapsing upon the OFF-alpha ganglion cell have the morphology of A19 amacrine cells (Fig. 18, A19, arrows). They are large-field with large-diameter dendrites tapering to fine, long processes covering 500 μm or more of field.

Like most wide-field amacrine cells of the mammalian retin, A19 cells are thought to be GABAergic and ON-OFF in physiology.

They run in lower sublamina a in S2 co-stratifying with OFF-centre alpha cell dendrites. A19 cells have been characterized, by their content of neurotubules and their gap junctions to other A19s. Moreover, HRP-injected A19 cells have been seen to run along and across large diameter alpha cell dendrites in sublamina a making intermittent synapses.

Amongst the many wide-field amacrine synapses upon OFF centre alpha cells, some are probably from acetylcholine containing or "starburst" amacrine cells (Fig. 18, ACh, stippled cell). Vardi and co-workers (1989) showed by lucifer filling alpha ganglion cells and immunocytochemically staining cholinergic amacrine cells, that the latter amacrine's distal varicose dendrites run in bundles and loops along and across alpha cell dendrites in strata 2 and 3 of the IPL having ample opportunity for synaptic interaction. There is also convincing evidence for physiological effects of acetylcholine upon Y ganglion cells and evidence that alpha cell dendrites contain ACh receptor sites. Thus it seems probable that the two cholinergic amacrine cell types contributing to a and b sublaminae in cat retina are providing synaptic inputs to OFF- and ON-centre alpha ganglion cells respectively.

Primate Ganglion Cells

There are at least 18 different morphological types of ganglion cell in the human retina. All but three of these correspond to cells in the cat retina and have, where possible, been given the same G number.

Some of the cat types have not been seen in human retina but undoubtedly they are there. Special names are reserved for P and M ganglion cells that are specific to the human high visual acuity and colour processing systems.

They are thought to be specialized evolutionary correlates of beta and alpha cells respectively and comprise the major input to the geniculo-striate areas. Thus P cells are considered to be mostly midget ganglion cells and equivalent to beta cells of cat, and project to the Parvocellular layers of the LGN. M cells, on the other hand, probably equivalent to alpha cells of cat and project to the Magnocellular layers of the LGN (Fig. 19, below).

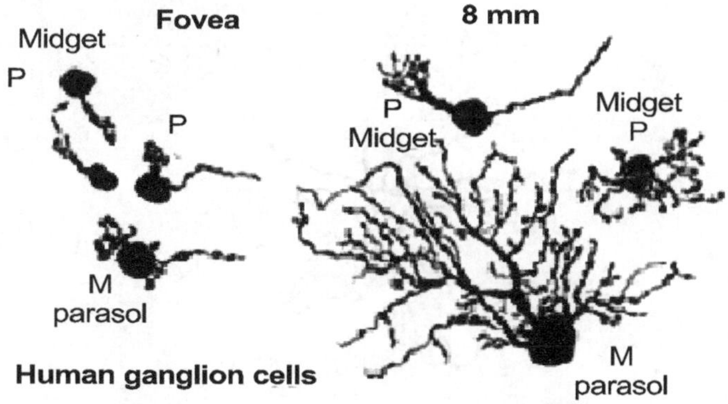

Fig. 19. Ganglion cell types involved in geniculate-striate pathways of human retina

Polyak (1941) first described the midget ganglion cells and considered them to form the majority of the ganglion cell types of monkey central retina. More recent retrograde staining studies indicate that ganglion cells with medium-size cell bodies and the smallest dendritic trees, putting them into a "midget" category of cell, comprise 80% of the ganglion cell population in the monkey retina, and these cells project to the parvocellular layers of the LGN.

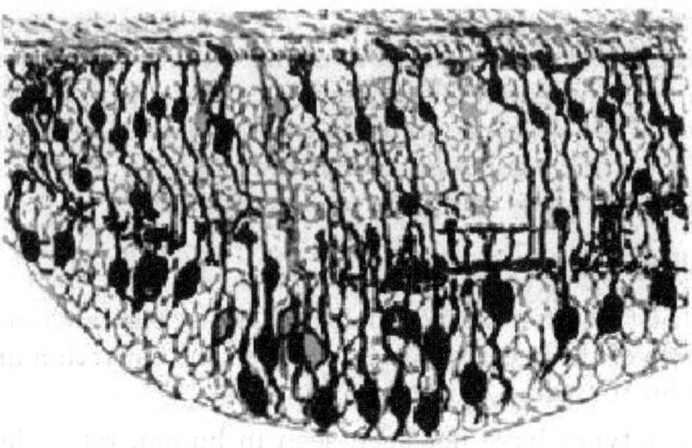

Fig. 19.1. The commonest ganglion cells types of the primate fovea are the P cells or midget ganglion cells and the bistratified blue-on/yellow-off ganglion cell.

The question then arises as to how many cell types correspond to these parvocellular projecting ganglion cells. Polyak (1991) describes midget ganglion cells quite clearly as the ganglion cells with the smallest dendritic trees never much exceeding the diameter of their cell body which is 9-12 μm. It is clear that the cells of the fovea meet this criterion (Figure 19). However outside of the fovea and beyond 3mm of eccentricity the midget ganglion cells can have dendritic tree sizes as large as 100 μm.

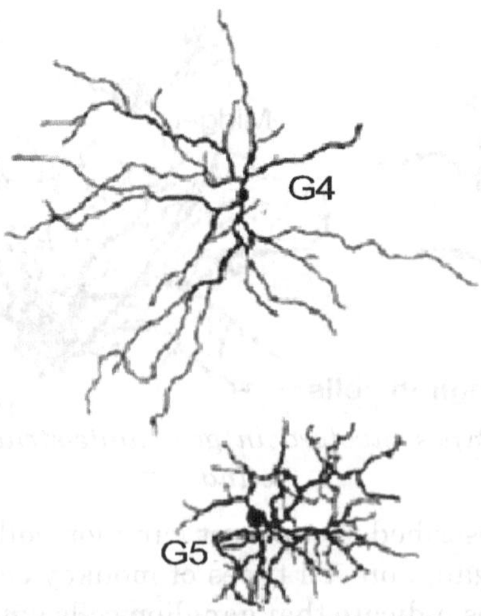

Fig. 20. G4 and G5 ganglion cells of human retina

Physiologically we suspect by virtue of their being single-cone-connected via their exclusive connection to single midget bipolar cells in the foveal area that P cells are the opponent chromatic units, or Type 1 cells of Weisel and Hubel, of the LGN. Outside of the fovea where they clearly are receiving input from more than one midget bipolar cell by virtue of their increased dendritic tree size midget ganglion cells or P cells may become broader band.

We suspect that the P cells always receive input through midget bipolar cells but probably from more than one midget bipolar cell. In peripheral retina the P cell dendritic trees become large enough for axons of 3-5 midget bipolar cells to have input. It remains to be seen whether chromatic specificity can remain under these conditions.

There is general agreement that the magnocellular-projecting ganglion cells are the parasol cells that Polyak described (1941). Although some of Polyak's giant cells are by cell body size and dendritic tree size characteristics undoubtedly the equivalent of Golgi-stained M cells in far peripheral retina and of the large-bodied, multibranched reduced silver stained cells of Silveira and Perry (1991).

The other ganglion cells in human that are not in the geniculo-striate projections correspond to some of those retrogradely-filled in Perry and co-workers studies of cells projecting to the midbrain (Figs. 20-23). Two of these varieties in macaque retina may be G22 and G23 cells (Fig. 23). The G19 is equivalent to the cell in cat retina that has been stained by intracellular dye-injection and is apparently monoamine accumulating (Fig. 22). Ganglion cells with complex receptive field characteristics such as directional selectivity (DS) may be present in human retina too. Thus G11 (Fig. 24) is a bistratified cell that could correspond to the rabbit ON-OFF DS cell and G22 (Fig. 23) to an ON directionally selective cell.

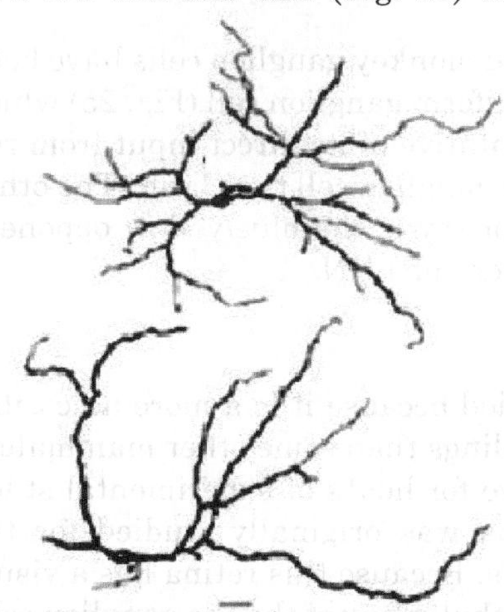

Fig. 21. G20 and G21 cells of human retina

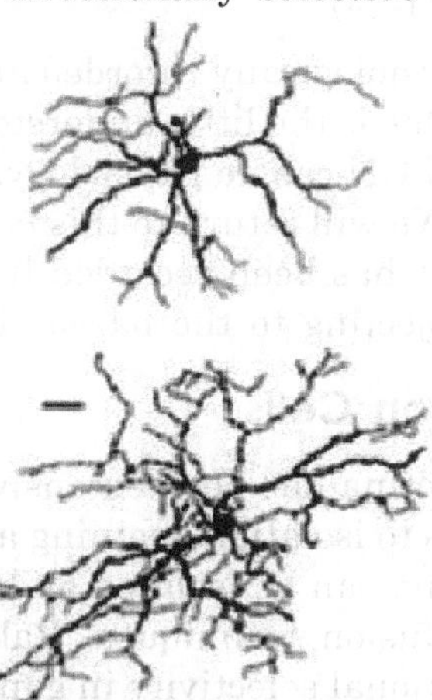

Fig. 22. G17 and G19 cells of human retina

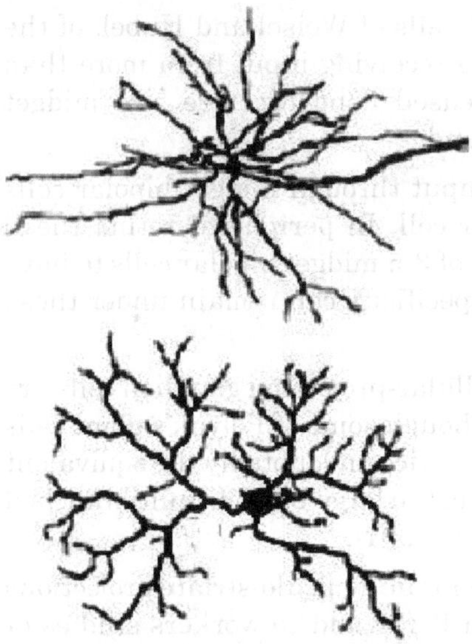

Fig. 23. G22 and G23 cells of human retina

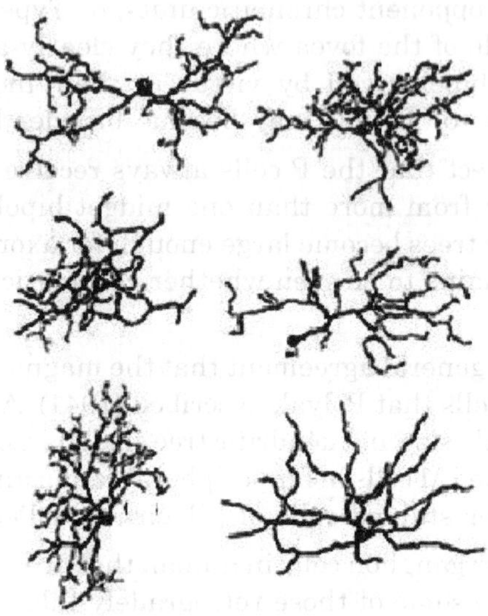

Fig. 24. Smaller field ganglion cells of human retina

Ganglion cells selective for orientation and uniformity could correspond to G8 (Fig. 24) and G17 (Fig. 22) respectively, in human retina. Several of the concentric sluggish cells of the rabbit are also seen in human Golgi-stained counterparts, for example G3 and G20 could be an ON-centre varieties and G7 and G4 OFF-centre varieties (Figs. 21-24).

Only two physiologically recorded and stained monkey ganglion cells have been reported so far. One is the little understood biplexiform ganglion cell (Fig. 25) which gives a sustained ON-centre physiology, representative of its direct input from rod photoreceptors. We will return to this fascinating ganglion cell type later. The other ganglion cell that has been recorded from in monkey is the blue/yellow opponent ganglion cell projecting to the parvocellular layers of LGN.

Rabbit Ganglion Cells

The rabbit retina has been extensively studied because it is a more accessible retina in respects to isolation, staining and recordings than some other mammalian retinas. The retina can be isolated and kept alive for hours of experimental study with simple perfusion techniques. Rabbit retina was originally studied for the property of directional selectivity in ganglion cells. Because this retina has a visual streak specialization, it was found quite early on that many of the the ganglion cells were sensitive to motion and direction of the movement of visual stimuli.

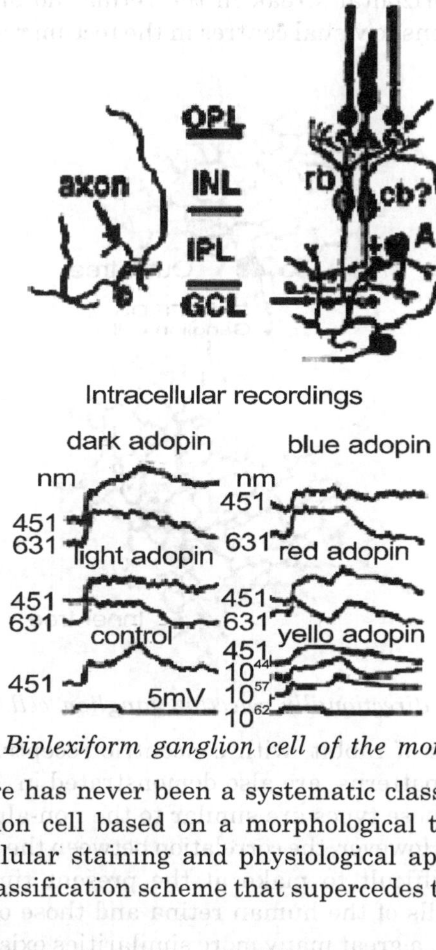

Fig. 25. Biplexiform ganglion cell of the monkey retina

Interestingly there has never been a systematic classification of the different types of rabbit ganglion cell based on a morphological technique like the Golgi. However, the intracellular staining and physiological approaches have yielded a very nice functional classification scheme that supercedes the need for a Golgi study nowadays.

Ganglion cells equivalent to ON and OFF-alpha types in cat are also seen in rabbit. They have both a similar physiology and dendritic field size and shape although in rabbit their cell bodies may not be quite as large in size as in cat retina. equivalent beta cell types in rabbit are a little more equivocal. Cells that respond with the linear, brisk sustained concentric receptive field organization that come in ON and OFF-centre pairs, are seen in rabbit and their appearance is rather similar to peripheral beta cells of cat. They have, however, smaller cell body sizes, again like the alpha cell comparison. Because no cell with the exact same morphology as a central beta cell, even on the visual streak of rabbit retina, has been seen some suggest that there is no real equivalent of the beta cell in rabbit. The question remains an open one, because the rabbit visual system has a different organization than the cats, with a horizontal streak in the retina, no binocular overlap of the visual fields and projections to visual centres in the tectum rather than the geniculo-striate.

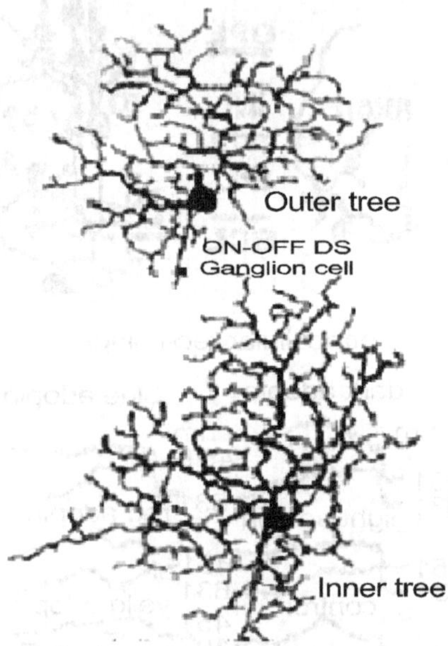

Fig. 26. ON-OFF directionally selective ganglion cell of rabbit retina

Other ganglion cells of rabbit, with concentric receptive field physiology but more sluggish response patterns, are also demonstrated in Amthor and co-authors work (1989a). Many of these types are similar to the non-alpha/non-beta cell types of the cat in morphology. However, the correlation between the ganglion cell physiology of the two species is difficult to make at the present time. Some comparisons between the ganglion cells of the human retina and those of the rabbit have been attempted. Interestingly, a great many more similarities exist between ganglion cell types of the ground squirrel and rabbit retinas.

The more unusual feature detectors of the rabbit retina are the motion sensitive, orientation and direction sensitive ganglion cells. The bistratified ON-OFF directional selective (DS) ganglion cells of the rabbit retina have attracted a lot of attention mainly because many aspects of the circuitry have been elucidated (Fig. 26). For example the acetylcholine containing amacrine cells, also called starburst cells, are the major input cells. The effects of antagonists of acetylcholine have been known for a long time to eliminate directional selectivity in the rabbit retina, and amacrines containing GABA are also known to be involved. The fascinating aspect of the ON-OFF directionally selective ganglion cells of rabbit is that they are bistratified and have the two layers of their dendrites oriented orthogonal to each other. The orientation of the outer dendritic tree (i.e. the OFF layer of dendrites) is favoured to be with the preferred direction of the cells response to a moving target

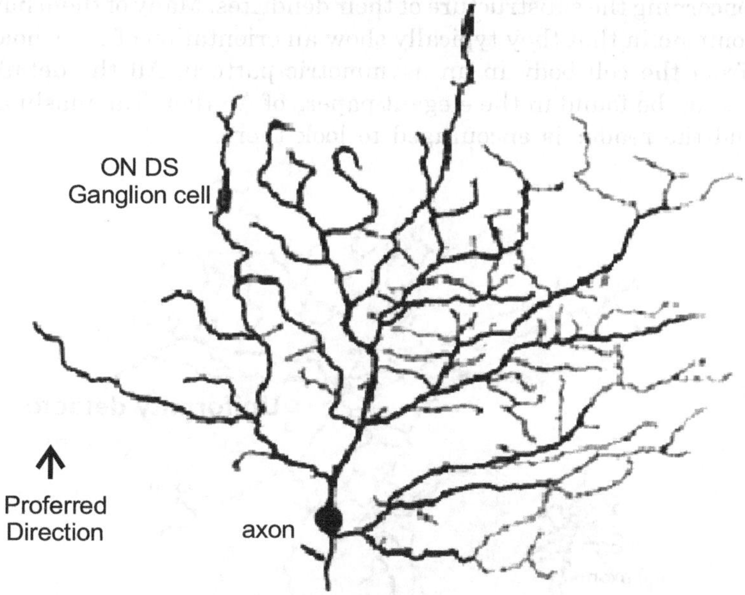

Fig. 27. ON-directionally selective ganglion cell of rabbit retina

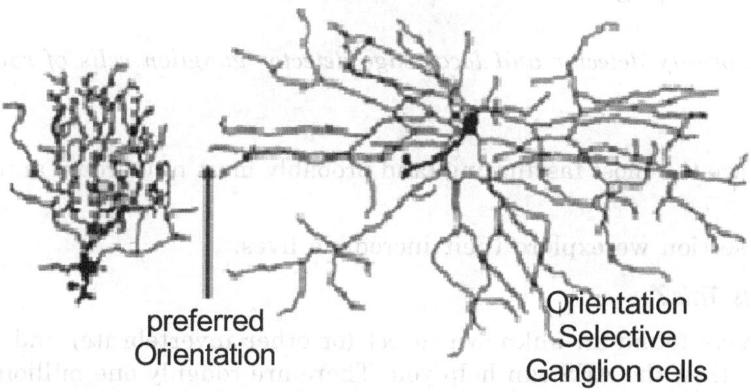

Fig. 28. Direction and orientation selective ganglion cells of rabbit retina

The other type of directionally selective cell in rabbit is the ON-centre type and its single layered dendritic tree in sublamina b (ON layer of the IPL) is not as obviously oriented with the preferred direction (Fig. 27). Other ganglion cell types that are prominent and well characterized in Amthor and coworkers studies (1989b) in the rabbit retina, include orientation selective, local edge detectors and uniformity detectors (suppressed by contrast) (Figs. 28 and 29). It is particularly interesting that each of the different types of ganglion cells have a distinctly different morphology, primarily concerning the substructure of their dendrites. Many of them have another feature in common in that they typically show an orientation of the whole dendritic tree away from the cell body in an asymmetric pattern. All the details of these morphologies can be found in the elegant papers of Amthor, Takahashi and Oyster (1989a,b) and the reader is encouraged to look there.

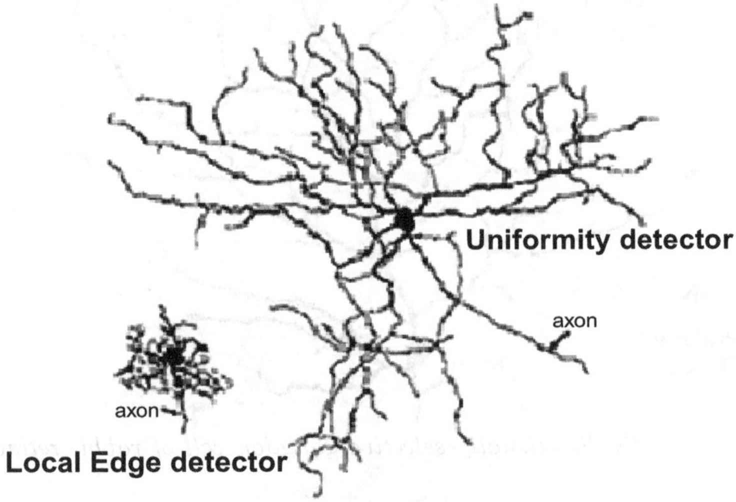

Fig. 29. Uniformity detector and local edge detector ganglion cells of rabbit retina

Insects

Insects are the most fascinating (and probably most numerous) animal on the planet.

In this section we explore their incredible lives.

What Bug is This?

If you have found an unknown insect (or other invertebrate) and would like to identify it then the AES can help you. There are roughly one million described species of insect and experts estimate that there are probably another five million species that are yet to be identified. If you combine that total with the millions of species of invertebrate that are not insects you'll realise that no one person can ever be an expert in any more than a relatively small area.

With that in mind the society provides information and two different identification services, one is for members of the society and the other is for non-members.

Introduction to Arthropods

Arthropods are animals that lack a backbone. Such animals are said to be invertebrates. The invertebrates includes animals such as snails (molluscs), crabs (crustaceans) and, of course, insects.

The arthropod body is supported by an hard, outer layer. This acts as an external skeleton-or exoskeleton. The body muscles are attached to the inside of this exoskeleton. It is made up of rings (called segments) which are attached to one another by flexible membranes.

This allows movement of the body to take place. The name arthropod refers to their jointed limbs, which are their chief characteristic. (Arthros = joint, Podium = foot). There are various types of arthropod and they are classified in to groups based on common features. With a little care they may be easily and quickly recognised in most situations:-

Centipedes

Centipedes (cent = 100, pedes = feet) do not have 100 legs, they usually have 40- 50. There is a distinct head, but the rest of the body shows a series of similar segments. Each of these segments generally has one pair of walking legs.

Centipedes are sometimes grouped together with Millipedes as Myriapods but they are more usually considered as a separate group called Chilopoda.

They are carnivorous (meaning they eat other animals), possessing a pair of poison jaws. The large tropical forms can reach 30cm in length and are some are quite dangerous. In general they show retiring habits, living in out of the way places such as under logs and stones.

General Structure of a Centipede

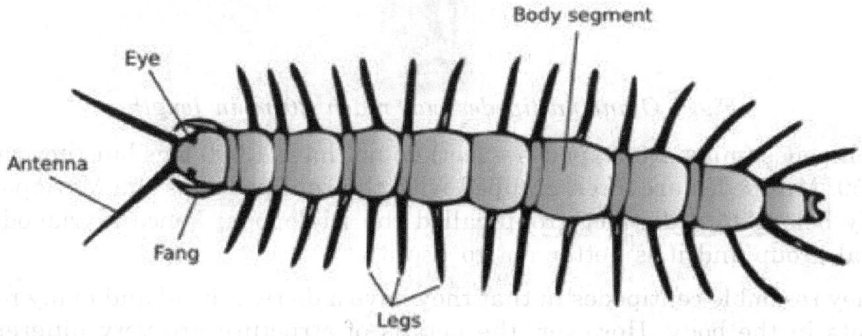

Note that there is one pair of legs per body segment as shown by the diagram above and the cross section and photograph below.

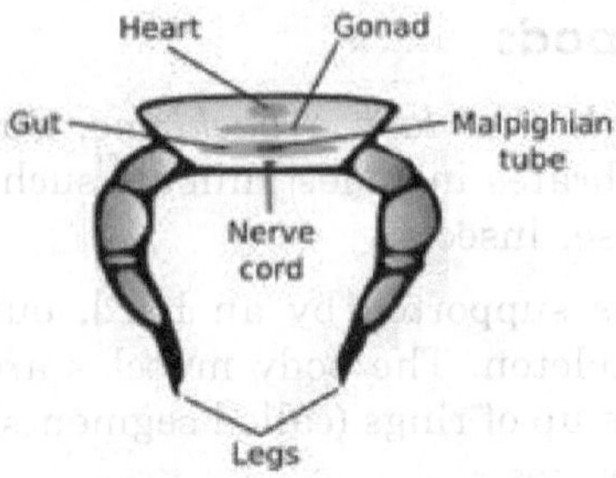

An illustration of a cross section of a centipede body segment.

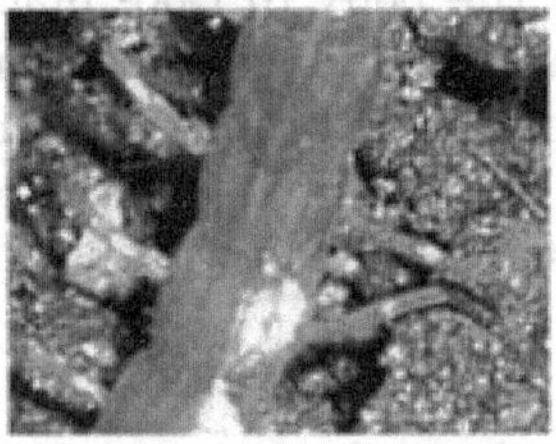

Fig. : Close up photograph of centipede body segments showing one pair of legs per segment.

Millipedes

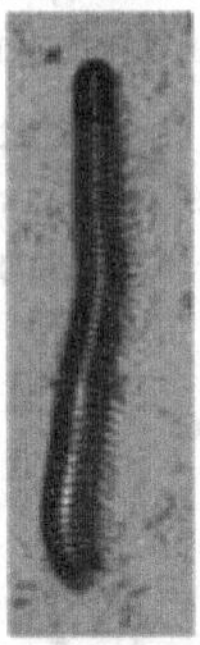

Fig. : Giant millipedes can reach 30cm in length.

Millipedes (mil = 1,000, pedes = feet) do not have 1,000 legs but they can have over 250. Millipedes are often grouped with the centipedes as the Myriapoda, but actually belong to a distinct group called the Diplopoda; hence Myriapoda is an artificial group and it is better not to use it.

They resemble centipedes in that they have a distinct head and many repeated segments in the body. However, the details of structure are very different. The easiest way of recognising them is to look at the legs. Most segments of the body appear to have two pairs of legs. The body is also more rounded in cross section, whereas the centipedes are more flattened.

General Structure of a Millipede

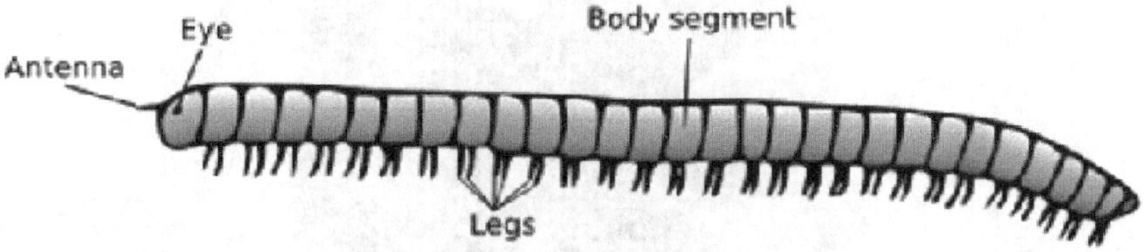

Note that there are two pairs of legs per body segment as shown by the diagram above and the cross section and photograph below.

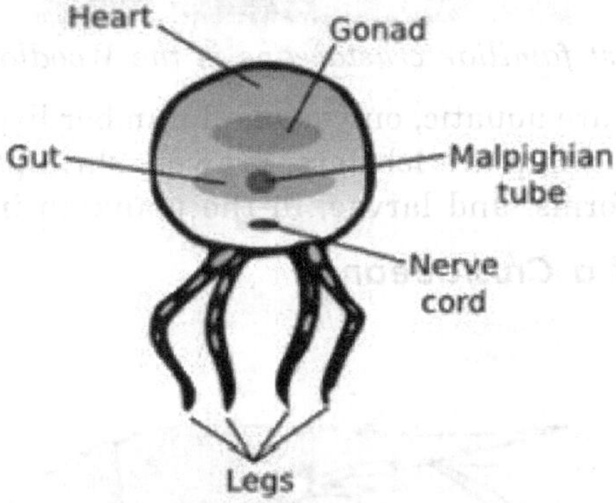

Fig. : An illustration of a cross section of a millipede body segment.

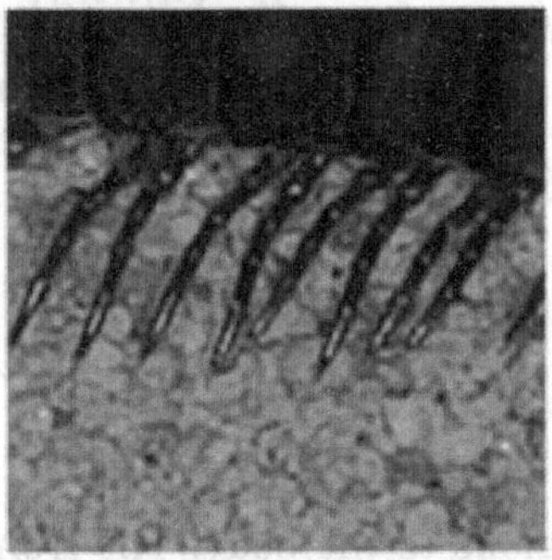

Fig. : Close up photograph of millipede body segments showing two pairs of legs per segment.

Crustaceans

Crustaceans have three regions to the body (head, thorax and abdomen). In some forms these are not clear in the adult. They usually have a number of pairs of legs on the thorax. Some have a small number- e.g. crabs have four pairs. Generally, the abdomen also has limbs which are often involved in locomotion.

Fig. : One of the most familiar crustaceans is the Woodlouse, Porcellio scaber.

Most Crustaceans are aquatic, only a small number live on land, e.g. woodlice. Other members of the group are lobsters, prawns, shrimps and barnacles. There are also many small forms, and larvae, in the plankton in water.

General Structure of a Crustacean

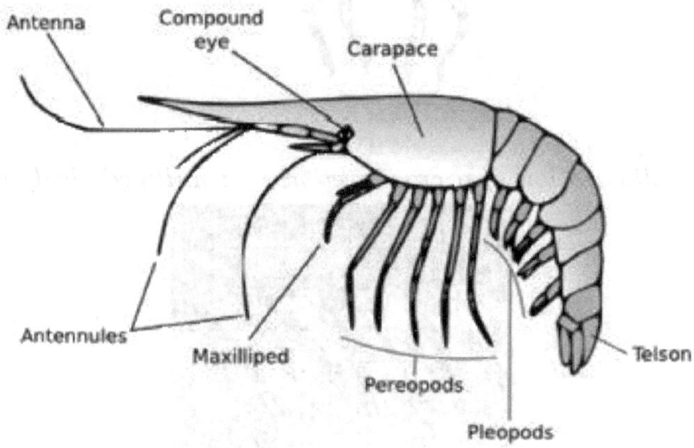

Arachnids

Arachnids have two regions to the body. The front (or anterior) part is called the *prosoma*. The rear (posterior) part is called the *opisthosoma*. There are four pairs of walking legs on the prosoma in adult arachnids (but larval mites only have three pairs).

Arachnids generally feed by sucking fluids from the bodies of other organisms. Hence they tend to be predatory or parasitic, though there are exceptions. The group includes:

- spiders
- ticks

- mites
- scorpions
- pseudoscorpions
- kingcrabs.

*Fig. : A common sight in British gardens, a Wolf spider (*Pardosa *sp.).*

General Structure of An Arachnid

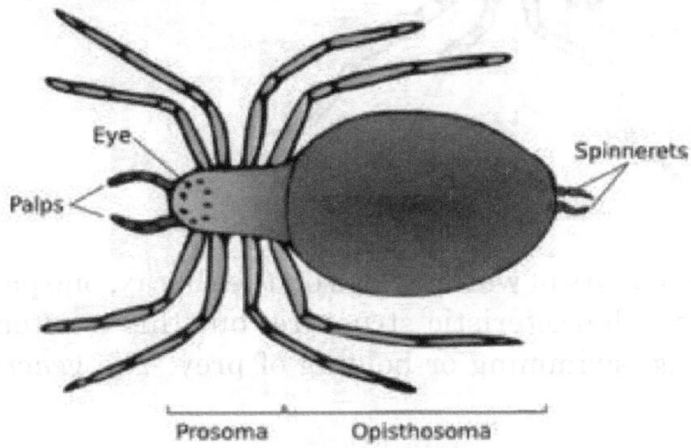

Insects

A Blue Morpho butterfly (*Morpho Sp.*).

Insects are the most numerous life form on the planet (in terms of number of species). Approximately seven out of every eight living species are insects. Insects show a number of characteristics, the three by which they are most easily recognised are:-

1. The body is divided into three distinct regions-head, thorax, and abdomen. Each region is further divided into segments. Typically, there are six in the head, three in the thorax and eleven in the abdomen. In the more advanced insects segments may become fused together, particularly in the abdomen. The structure of a generalised insect is illustrated below, but insects are a very diverse group and have evolved many different forms:-

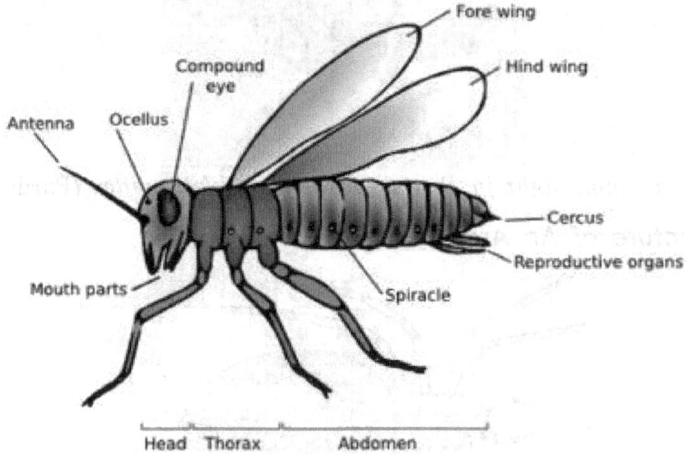

2. There are three pairs of walking legs on the thorax, one pair to each segment. The legs show a very characteristic structure, but this is often modified to fulfil a variety of tasks, e.g. swimming or holding of prey. The *generalised* structure of an insect leg:-

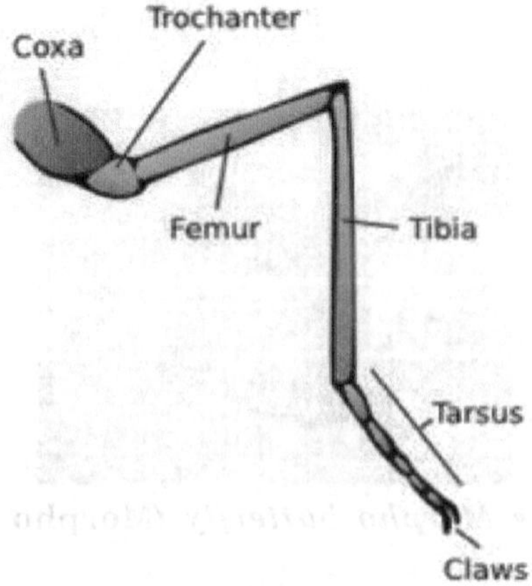

3. The adults of most insects show two pairs of wings, one pair on each of segments two and three. The wings are supported by a series of veins, the pattern of veins being important in classification. The naming of the veins and regions on an insect wing:-

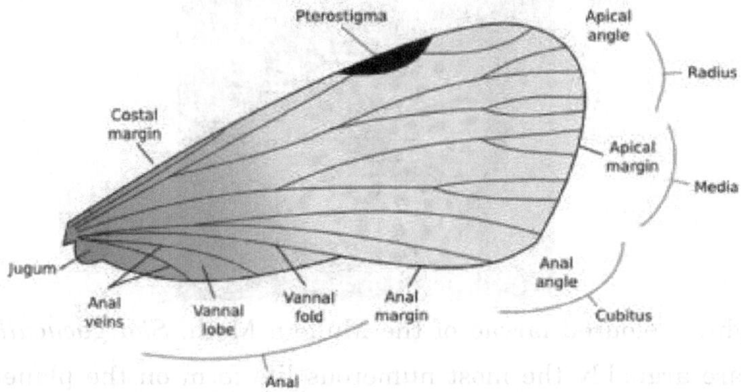

Respiration

Insects 'breathe' through a system of branching tubes, the tracheae. Oxygen and carbon dioxide move along these by a process called diffusion. The tracheae open on the surface of the body at special pores, the spiracles. The insect may control the size of these pores. It is this method of 'breathing' which stops insects from getting very large. The insect body cannot get bigger than a diameter of about three centimetres. Above this size diffusion of oxygen into the body tissues becomes too inefficient for the insect to live.

Vision

The head bears a pair of compound eyes. These consist of a number of individual 'eyes', each of which produces a separate image. Hence the overall picture that the insect sees is made up of a series of dots. This is rather like a television picture, but with much poorer sharpness. This type of eye is very good at judging distance and movement. Hence insects which are active predators, such as dragonflies, have very well developed eyes.

Antennae

The antennae (or 'feelers') are mainly organs of smell and taste. They do have other functions in certain insects, e.g. they may be used to detect air currents. Hence the word 'feeler' does not describe their function and should not be used. They are made up of individual 'segments' that are attached to one another by flexible membranes. There are muscles inside the antennae that allow them to move. Antennae come in a variety of forms, and may be different in male and female, e.g. in chironomid midges. The antennae of male chironomid midges are 'bushy' because smell is very important in finding a female to mate with.

Insects fact files

The brightly coloured larvae of the Mullein Moth, *Shargacucullia verbasci.*

Insects are arguably the most numerous life form on the planet in terms of number of species. There are roughly one million described species of insect and experts estimate that there are probably between five to six million species yet to be identified. The structure of insects has allowed them to exploit niches in almost all the habitats of the world. In this section we examine insects in more detail, focusing not only on their body structure but life cycles and providing fact files on all the major groups (Orders) of insects. There is also a general overview of the structure of insects in the What bug is this? section.

Insect Antennae

The large filiform antennae of a solitary bee.

The antennae are often called 'feelers' because the insect waves them around. This is a wrong name because they are not only used for touch. The antennae are actually the insects 'nose'-they are used for the sense of smell.

The paired antennae are made up of a number of individual joints. This means they can be very mobile. The basic form of antenna is filiform. In this type there are many segments that are more or less equal in size. Filiform antennae are seen in a wide variety of groups, such as Dragonflies, Grasshoppers and Crickets, Book Lice, Biting Lice, Scorpion Flies and Beetles. The length and number of joints varies much between them.

Filiform Antennae

This is the most basic form of insect antennae.

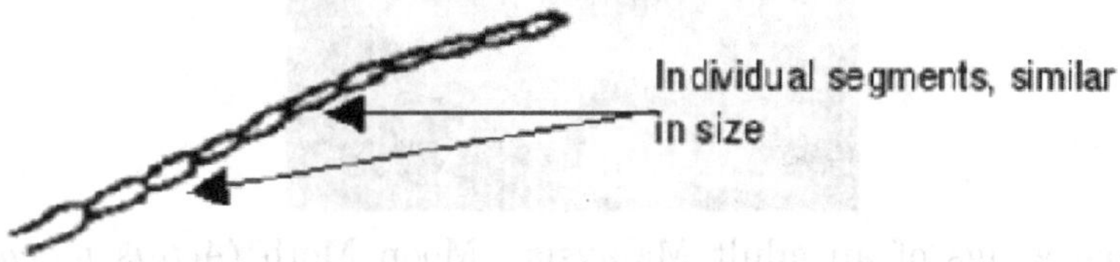

Fig. : Illustration of filiform antennae.

This basic structure is modified in a wide variety of ways. This means that a number of different types may be recognised. The main ones are as follows:-

1. Setaceous-There are many joints. The antenna tapers gradually from the base to the tip e.g. Bristletails, Cockroaches, Mayflies, Stoneflies and Caddisflies.
2. Moniliform-The round segments make the antenna look like a string of beads e.g. Beetles.
3. Serrate-the segments are angled on one side giving the appearance of a saw edge e.g. Beetles.
4. Pectinate-The segments are longer on one side. This gives the appearance of a comb e.g. Sawflies (related to wasps) and Beetles.
5. Clavate-the segments become wider towards the tip of the antenna. This may be gradual along its length, or a sudden increase and therefore mainly affecting the last few joints and giving the appearance of a club e.g. Butterflies & Moths and Beetles.
6. Lamellate-the segments towards the end are flattened and plate-like. This gives the appearance of a fan e.g. Beetles
7. Geniculate-there is an abrupt bend or elbow part of the way along the antenna e.g. Ants and Beetles.
8. Plumose-the segments each have a number of fine thread-like branches. This gives the appearance of a feather e.g. Flies.

The antennae of the Lepidoptera, Neuroptera, Hymenoptera and Coleoptera are very variable. These groups show examples of a number of different antenna types. All the examples quoted are from insect groups with British representatives. Where groups have no British representatives.

Insect Wings

The large wings of an adult Malaysian Moon Moth (*Actias maenas*). Moon moths are a type of Silkmoth. Apart from the vertebrates, insects are the only other group of animals to have successfully evolved flight. The wings are composed of two membranes of the cuticle pressed together and supported by a series of veins. The pattern of veins, the venation, is not haphazard but very regular, though it does show modifications. These are very useful in identification. The wings are more or less triangular in form and certain regions may be recognised:-

Regions of the Wing

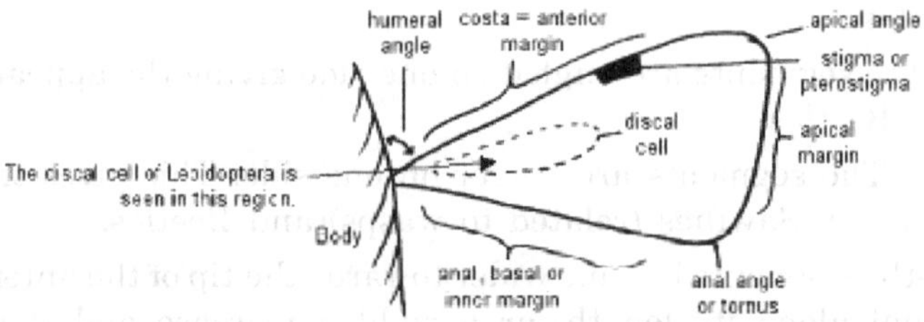

Fig. : Illustration of the main regions of the insect wing.

Fore Wing

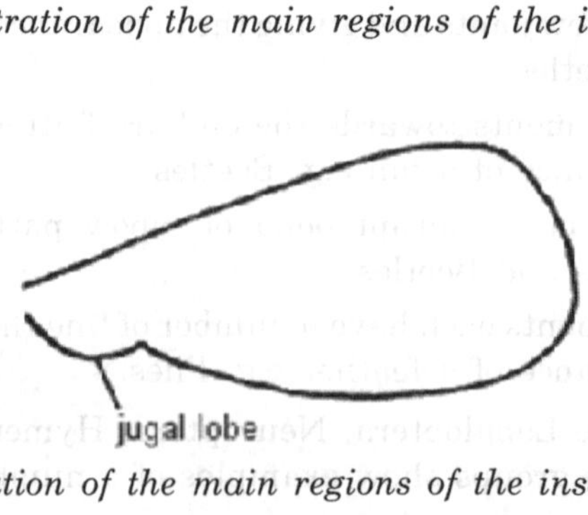

Fig. : Illustration of the main regions of the insect fore wing.

Hind Wing

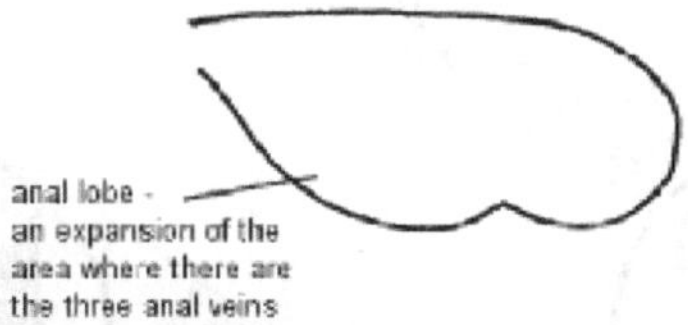

Fig. : Illustration of the main regions of the insect hind wing.

The basic pattern of longitudinal veins, the venation, have been shown in a previous article. This pattern may be modified in a variety of ways, such as fusion of veins, loss of veins or even the development of additional ones. Primitively these main veins are connected by a series of cross veins which are seen clearly in fossil forms as well as present day Dragonflies & Damselflies and Mayflies

Dragonfly Wing showing the series of cross-veins

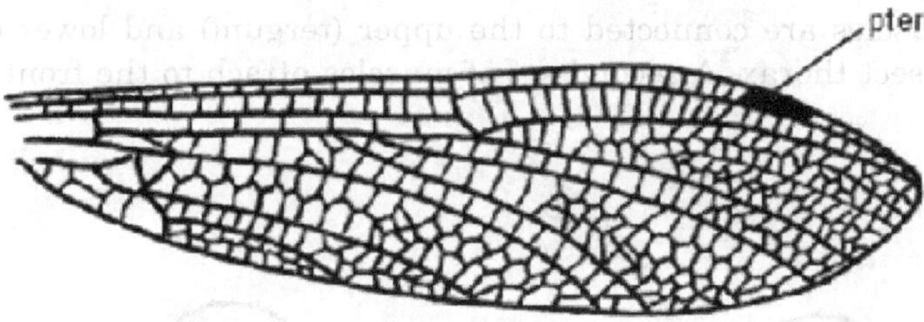

Fig. : Illustration of the cross veins on a Dragonfly wing.

In most insect orders these cross veins are reduced or absent. Primitively the pairs of wings beat independently of one another.

This is rather inefficient, the effectiveness of the wing beat may be increased by the wings acting together. A number of different methods of coupling the wings to achieve this have been developed. The jugal area of the fore-wing is involved in some of these.

Flight Muscles

Insects have one of two different arrangements of muscles used to flap their wings:-

Direct Flight Muscles

Direct flight muscles are found in insects such as dragonflies and cockroaches. The wings pivot up and down around a single pivot point. The wings are raised by a contraction of muscles attached to the base of the wing inside (toward the middle of the insect) the pivot point. The wings are then brought down by a contraction of muscles that attach to the wing outside of the pivot point.

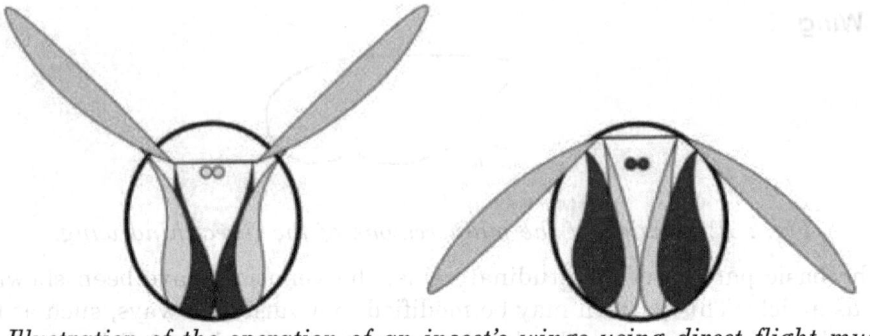

Fig. : Illustration of the operation of an insect's wings using direct flight muscles. The darker muscles are those in the process of contracting.

Indirect Flight Muscles

Indirect flight muscles are found in more *advanced* insects such as true flies. Indirect flight muscles are connected to the upper (tergum) and lower (sternum) surfaces of the insect thorax. A second set of muscles attach to the front and back of the thorax.

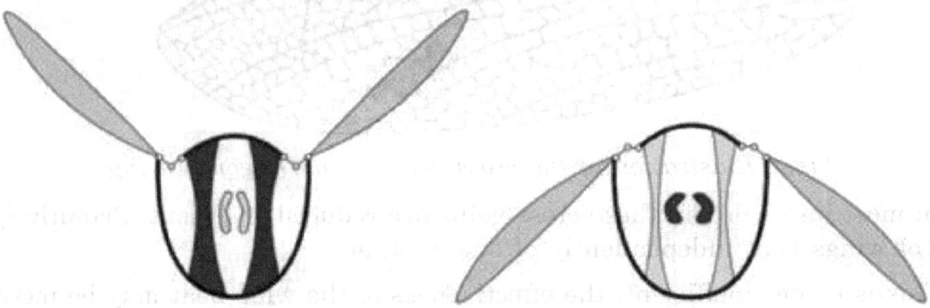

The wings are raised by the muscles attached to the upper and lower surface of the thorax contracting. This brings the top surface of the thorax down and, along with it, the base of the wings. As a result the wing tips pivot upwards. The wings are then lowered by a contraction of the muscles attached to the front and rear of the thorax. This forces the upper surface of the thorax to raise and the wings pivot downwards.Illustration of the operation of an insect's wings using indirect flight muscles. The darker muscles are those in the process of contracting.

Synchronous and Asynchronous Muscle

Insects that beat their wings less than one hundred times a second use synchronous muscle. Synchronous muscle is a type of muscle that contracts once for every nerve impulse.

Insects that beat their wings more rapidly use asynchronous muscle. This is a type of muscle that contracts more than once per nerve impulse. This is achieved by the muscle being stimulated to contract again by a release in tension in the muscle. This can happen more rapidly than through simple nerve stimulation alone.

Insect Respiration

Fig. : The spiracles are clearly visible as black dots on the side of each body segment of this beetle larva.

Insects take in Oxygen and expel Carbon Dioxide using a series of internal air tubes, the tracheae. These pass fine branches, the tracheoles, to all parts of the body. The cuticle lines the tracheae. These linings are shed with the rest of the cuticle when the insect moults. There is a main pair of lateral tracheae running the length of the body, one on each side. Some insects show two such pairs, one dorsal, the other ventral.

From the main trunks, tracheoles pass throughout the body. The tracheae open to the air at the paired spiracles. There is one pair per segment in the thorax and also in the first nine abdominal segments. This arrangement does vary with different species. In most species the spiracles on the first segment of the thorax are missing. The pair on the ninth abdominal segment is also missing.

Diagram of Part of the Tracheal System of a Cockroach

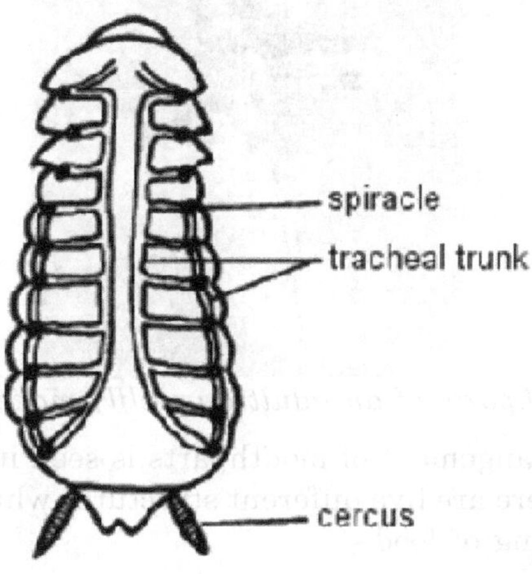

The tracheoles end within the body cells. Gases move by diffusion within the tracheal system. When the insect is less active the ends of the tracheoles contain fluid. It is where the fluid and gas meet (= the fluid/gas interface), that exchange of gases occurs (oxygen is taken up, carbon dioxide is given off). As activity increases, the fluid is removed from the tracheoles. This means that the exchange of gases occurs nearer the cells. In the extreme case of fatigued flight muscle, the exchange interface lies within the muscle cells.

Gas/Water Interface in the Tracheole

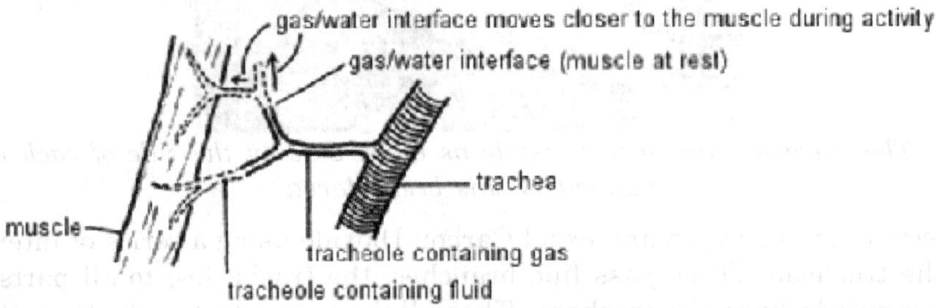

Diffusion alone does not meet the respiratory demands of the animal. Gases are moved, to a large extent, by pumping actions of the body segments.

The respiratory system of an insect is very efficient for small organisms. As body size increases, the efficiency decreases. When body diameter exceeds about 3cm, the respiratory needs cannot be met. Hence it is the respiratory system of insects which restricts their body size. Note that no transport system, such as blood, is involved in moving oxygen or carbon dioxide around the body.

Insect Mouthparts

Fig. : The biting mouthparts of an adult damselfly make short work of an aphid.

The 'primitive' arrangement of mouthparts is seen in the cockroach-here they are used for biting. There are five different structures which are used for the initial collection and processing of food:-

What are the different parts for?

1. Labrum-a cover which may be loosely referred to as the upper lip.
2. Mandibles-hard, powerful cutting jaws.
3. Maxillae-'pincers' which are less powerful than the mandibles. They are used to steady and manipulate the food. They have a five segmented palp which is sensory and often concerned with taste.
4. Labium-the lower cover, often referred to as the lower lip. It actually represents the fused pair of ancestral second maxillae. They have a three segmented palp which is also sensory.
5. Hypopharynx-a tongue-like structure in the floor of the mouth. The salivary glands discharge saliva through it.

This system remains little changed in all insects which chew their food, both larvae and adults. Where specialised food sources have been exploited, the mouthparts are modified, sometimes very considerably, so that the food may be obtained satisfactorily. Insect mouth parts show many cases of parallel evolution, the same end being independently achieved along similar, but not identical lines. Many insects take in liquid food. This is facilitated by the development of a sucking' arrangement from the mouthparts.

Mouthpart Modifications

1. Honey bee-the mandibles are very small and suitable for moulding wax, the labium is curved downwards and inwards forming a tube used for sucking up nectar.
2. Butterflies and Moths-the mandibles have disappeared altogether. The maxillae are elongate, channelled along their surfaces and held together by hooks and spines to form a sucking tube. This proboscis may be very long, when not in use is carried coiled up like a watch spring under the head.
3. Diving beetle larvae-have a piercing and sucking arrangement. The mandibles are each curved over to form an almost closed groove along their inner surface. The prey is caught and pierced by the mandibles. Digestive juices are pumped down the groove. The food is made into a liquid. This semi-digested soup is sucked back up again through the groove.
4. True bugs-these also have a piercing and sucking arrangement. The labium is elongated and acts as a sheath. This encloses the mandibles and maxillae which are modified as stylets for piercing. The maxillae have two tubes running along their length on the inside surface. 'Saliva' may be pumped down one of the tubes. This saliva makes the food into a liquid. The liquified food is sucked up the other tube.
5. House-fly-the proboscis shows a capillary mechanism. The labium is elongated and forms two lobes at the tip. These lobes have a series of fine tubes

(pseudotracheae). The hypopharynx runs down the proboscis and digestive juices pass down this onto the food. The food is made into a liquid by these juices. This liquid is then drawn up the pseudotracheae by capillary action.

Insect Life Cycles

Fig. : The pupa of the Large Copper butterfly (Lycaena dispar). Butterflies undergo complete metamorphosis.

The main disadvantage of the exoskeleton is that it can't expand with growth. In order for growth to take place, the exoskeleton must be shed and a new one formed. The new one will be soft at first, so the body must be expanded before the new one hardens. The organism can then grow to fill the space created before moulting becomes necessary again.

The process of moulting is called *ecdysis*, and the stage between successive moults an *instar*. Once adulthood has been reached, growth ceases and the adult nsect doesn't moult again. This means that the stages occurring before the adult are the ones in which growth occurs.

There are two different types of insect life cycle-*incomplete metamorphosis* and *complete metamorphosis*. (Metamorphosis means a change of form.)

Incomplete Metamorphosis: This is shown by the less highly developed insects. The life cycle shows only three stages:-

Egg-nymph-adult: The nymph resembles a miniature adult but is not able to produce young. This life cycle has the disadvantage that both nymph and adult often share the same food source. Therefore they can be in direct competition with one another for food. The advantage is that the vulnerable pupal (chrysalis) phase is avoided. The wings develop during the nymph stages as wing buds. These grow larger at each successive instar. They are fully formed at the final moult into adulthood.

Complete Metamorphosis: This is shown by the more highly developed insects. The life cycle shows four stages:-

Egg-larva-pupa-adult: The larva (or caterpillar) is the growth phase. It is generally very different from the adult. Usually the larva and adult use different food sources. Therefore they are not in direct competition. This is a distinct advantage as more individuals of the species can be fed. The pupa (or chrysalis) is a stage of internal reorganisation.

There are no visible signs on the outside of the body as to the activity within. Because of this the pupal phase used to be called the "resting" stage. During the pupal phase the internal organs are broken down, more or less forming a "soup". This "soup" then acts as food for special growth buds to develop. These form the adult body. When reorganisation is complete, the adult is ready to emerge. When outside conditions are suitable, the final moult occurs and the adult insect emerges. The wings develop inside the pupal case. Therefore this type of development is said to be endopterygote. Such a life cycle is shown by butterflies, moths, bees, wasps, true flies and many other insects, including beetles:-

Insect Overwintering

*Fig. : These caterpillars of the Large White butterfly (*Pieris brassicae*) only have to survive a mild frost. However, how do invertebrates survive the harsh winter cold?*

There is no doubt that winter is a difficult time if you are an invertebrate. Not only does the cold slow them down making them easier prey for hungry birds, but temperatures below zero can kill. To combat this insects and spiders have evolved lots of strategies to survive cold weather.

In places such as Canada and Central Europe night time temperatures can fall to as low as-40°C, yet insects survive this. We know they do because we see these places full of them again every summer. So how do they do it?

Aquatic Invertebrates

Water is an excellent insulator and, because the water does not lose heat as easily as the land, the temperature of the water stays quite constant. However, invertebrates living in water during winter have two things to worry about.

1. Will the water freeze?
2. If the water does freeze, will it all freeze or just the surface?

If the water doesn't freeze or only occasionally a thin layer freezes on the surface then you have no problems. Food is often plentiful. For example, many of the leaves that fell off the trees in autumn end up in streams and ponds and they provide an excellent food reserve.

Some invertebrates eat the leaves directly whilst others feed on the fungi and bacteria that rot the leaves down. These animals are called detritivores. Of course, all of these detritivores are food for carnivores so for many aquatic invertebrates, winter is a time of major growth.

Another reason that winter is a time for growth in aquatic invertebrates is oxygen. Aquatic invertebrates still breathe oxygen only they get the oxygen from the water not the air (unlike terrestrial invertebrates or ourselves). The water absorbs oxygen from the air and the colder the water the more oxygen it can absorb. This means invertebrates (and fish) have more oxygen available to them in cold water and can be more active.

However, if the surface of the water freezes over then air is not in contact with the water, so oxygen cannot be absorbed. Thick and persistent ice can mean that oxygen becomes very scarce.

If the pond or stream freezes solid then the invertebrates have the same problems as those on the land. Often though the bottom of a pond will not freeze and even if it does it is usually warmer than the land.

Terrestrial Invertebrates

The problems facing invertebrates over winter can be broken down into two categories:

Food: One problem during winter is a lack of food-most of the trees and shrubs have lost their leaves and there is little if any sap running in twigs. This means herbivorous insects (insects that feed on plants) have to find a way to live without food for months. To do this they enter a dormant (inactive) state like a deep sleep with all their metabolic (body) processes slowed down.

In winter this state is called *hibernation*. When insects use the same trick to avoid a hot dry period in summer (for example, in Africa or India) it is called *aestivation*. Of course, no herbivores means no food for the carnivores (animals that eat other animals) so the carnivorous insects and spiders also have to survive winter without food.

The Winter Moth (*Operophtera brumata*) is one of several moths that can fly even when temperatures approach 0°C.

Photograph by Olaf Leillinger licensed under Creative Commons.

Like everything in nature there are exceptions to these rules, and not all insects hibernate over winter. Some exceptions are the moths you occasionally see at night such as the Winter Moth (*Operophtera brumata*) and the December Moth (*Poecilocampa populi*) that can both fly at temperatures close to zero. Other exceptions are the many Collembola which don't seem to mind the cold and are often seen hopping around on the surface of snow.

A third exception are some Linyphiid spiders. About 9% of spiders in relatively mild areas remain active throughout winter. These are nearly all Linyphiids who can still make a web at temperatures as low as-1°C. They feed mainly on Collembola, but even they freeze and die if the temperature gets down to -7°C.

Cold Weather: How do terrestrial invertebrates survive the freezing conditions? There are several ways:

1. Avoid the cold weather
2. Protect yourself against the cold weather.

Avoid the Cold Weather: One way to avoid cold weather is to migrate to a warmer climate and return after winter. Perhaps the best example of this is the Monarch butterfly (*Danaus plexippus*) in North America.

Monarch butterflies migrate south from North America every year and overwinter in Mexico or in California, USA. For some individuals this migration is roughly 3,000 miles. In their overwintering sites the butterflies gather in trees in huge numbers. In spring the butterflies migrate back again.

Monarch butterflies (*Danaus plexippus*) gather in huge numbers in trees in Santa Cruz California to overwinter. Photograph by Mila Zinkova, used under GFDL

The second method of avoiding cold weather is to burrow down into the soil or leaf litter. Being in the soil is similar to being in water and is much warmer than being on the surface. Spending winter deep in the soil means that you not only avoid the cold but also the chilly winds and the beaks of hungry birds.

Many invertebrates overwinter in the soil. In areas where it snows regularly the snow can act like a giant blanket keeping away the worst of the winter weather.

Another favourite place to avoid both chilling winds and the probing beaks of birds is beneath the bark of trees. Occasionally the cold weather can still reach insects under bark and those in the soil.

Protect yourself against the cold weather: Some invertebrates use their body fluids to protect themselves against the cold. This protection comes in two forms:

1. Freeze tolerant invertebrates are those that can survive being frozen solid. Invertebrates survive the freezing by controlling where ice crystals form within their bodies. Although the ice crystals form inside the body of the invertebrate they do not damage the cells and organs of the animal. When the weather gets warmer the crystals melt and the invertebrate becomes active again.
2. Freeze intolerant invertebrates are those that use special "anti-freeze" chemicals to stop themselves freezing. These chemicals are called polyhydroxy alcohols and they work with other components of the invertebrate's body fluids (thermal hysteresis proteins) to prevent ice forming inside the insect.

Freeze intolerance is usually a strategy used by invertebrates found in cool to mildly cold climes. Freeze tolerance is used in really cold areas.

The "anti-freeze" chemicals are found in both freeze tolerant and freeze intolerant invertebrates.

The freeze tolerant insects use these chemicals to prevent ice crystals forming inside their cells whilst the haemolymph around them freezes. Freeze intolerant insects use the chemicals to try and prevent freezing altogether. However, if the temperature drops too low then they can't stop ice crystals forming inside their cells and this can kill both freeze tolerant and freeze intolerant invertebrates.

By stopping ice from forming an invertebrate can become "supercooled". Supercooling is when a liquid is cooled to a temperature below its freezing point and it does not freeze.

Freeze intolerant insects, those that rely on super cooling themselves without freezing, will freeze solid and die if the temperature drops below their supercooling point. In comparison, freeze tolerant insects still freeze solid but can survive this freezing.

The supercooling point is the temperature below which an animal is unable to stop itself freezing despite the chemicals in its body.

A good example of an insect that is both freeze tolerant and has a low supercooling point is the beetle *Pytho deplaratus*. It can supercool (remain unfrozen) to temperatures as cold as -54°C. If the temperature drops further it can still survive for a while because it is freeze tolerant and ice forms around its cells not within them. Two more tricks invertebrates use are to reduce the amount of water in their body and also empty their gut of food. Emptying your gut of food works both to reduce your overall water content and to remove a possible source of ice crystal formation.

The strategies used by insects and other invertebrates to survive winter are many and varied. Fortunately the strategies work and, when the weather gets warmer, the insects return in great numbers.

Insect Classification

Fig. : An adult male Small White butterfly (Pieris rapae), just one of two species of butterfly that are called Cabbage Whites.

People recognise things by giving them a name. This is also true for living things. Common ones have names from everyday language. For example, the Cabbage White Butterfly, or the hoverfly, or the dragonfly, or the ladybird. These names are fine for general everyday use, but they do have problems:-

1. There are two different types of butterfly that are called the Cabbage White.
2. There are over 250 different types of hoverfly in Britain.
3. The name dragonfly refers to a number of different species, not just to one.
4. In other parts of the World, ladybirds are not called ladybirds. For example, in North America ladybirds are called ladybugs.

There are other problems as well. One of these is that the name can be different in different parts of the country. For example, a woodlouse is called a *Jiggy Pig* in some parts of the country, and by other names elsewhere. Another problem is that animals that are rarely or never seen won't have an everyday name.

Several hundred years ago, as science was developing, scientists started to name and describe living things in Latin. Often the process became tedious and at times the descriptions became the name! The situation became very chaotic.

Eventually, a Swedish scientist called Carl von Linné tried to sort it all out. (The Latin version of his name is Carolus Linnaeus-he lived 1707-1778.) He devised a system using 2 names. One of these is the species name (in older writings it may be called the trivial name). This species name is more or less unique to that specific type of plant or animal. Also, he wanted his system to show how different things were related. He therefore grouped different species that he thought were related into a genus (plural genera). Hence, each different type was called a species and had two names:-

Genus Species

The genus name is written first and begins with a capital letter.

The specific (or species) name is written second. It does not have a capital letter. Sometimes the species is named after someone. In these cases the name may begin with a capital letter.

The words used for the names were generally in Latin, the language of science and learning at the time. To make it stand out, the name was underlined when written. When typesetting a book, the printer will put underlined words into *italics*. Hence, in a book, the scientific names are in *italics*.

The Latin for 2 is "bi". A Latin term for name is "nome". If these two words are put together, we have the word "binomial". Hence, the two-name system devised by von Linné became known as the "Binomial System".

He tried to show how different genera were related by grouping them into a family. A number of different families were grouped into an order, orders into class, classes into a phylum, and several phyla into kingdoms. The two main kingdoms are the plants and the animals This list of groups (or hierarchy) may be shown as follows:-

- Species-Many characteristics the same.
- Genus-The number of shared features (or characteristics) gets less as we go down the list.
- Family
- Order
- Class
- Phylum
- Kingdom-Few characteristics the same.

This system tries to show how the different types of living things are related, it is called a "Natural System of Classification". He published his ideas in his book "*Systema Naturae*" in 1735. This was revised a number of times. The tenth edition, published in 1758, is the most important. Modern classification in Zoology is based on this. Linnaeus was influenced a lot by the earlier writings of the English naturalist John Ray (1627-1705). In fact, much of Linnaeus' work is based on, and is an extension of, Ray's works. This is particularly true of the classification of plants. Linnaeus's ideas on plant classification were published in his work "Species Plantarum", published in 1753.

The Binomial name is followed by the name (or abbreviation) of the person who first described the species. This person is called the author.

The Binomial System is used by scientists all over the world. This means that the names are international. Therefore, if you write an article about *Pieris brassicae*, all scientists will know what species you are writing about. This would not be true if you wrote about the Cabbage White Butterfly.

Many people are "put off" by the scientific names. This is a shame because they are generally easy and many people use them without realising when they talk about plants. The best way is to use them from the beginning. It doesn't matter if you pronounce them wrongly, eventually you will learn the correct pronunciation. The important thing is to use them. You will probably have trouble at first. Do persevere as you will soon get the hang of them. Then you will wonder why you were wary of them in the first place!

Insect Groups (Orders)

Fig. : A pair of mating butterflies. Butterflies and moths belong to the order Lepidoptera.

The Insects (Class Insecta) are divided into a number of Orders. These are grouped together into two sub-classes called the Apterygota (*wingless* insects) and the Pterygota (*winged* insects)-for further information on Classes, Orders and Sub-classes see the Classification section. In addition to the Class Insecta there are three other classes of invertebrates that have six legs, these three classes are known as non-insect hexapods.

The Insects-Class Insecta

Each of the insect orders are listed below and, by following the links, information on each order is provided (main characteristics, life cycle and other interesting facts) along with links to other Web sites containing further information:-

- [Subclass] Apterygota
  * Archaeognatha (Order: Microcoryphia)
  * Three-pronged bristletails (Order: Thysanura)
- [Subclass] Pterygota
  * [Division] Exopterygota
  * Biting and Sucking lice (Order: Phthiraptera)
  * Booklice and Barklice (Order: Psocoptera)
  * Cockroaches (Order: Blattodea)
  * Dragonflies and Damselflies (Order: Odonata)
  * Earwigs (Order: Dermaptera)
  * Grasshoppers and Crickets (Order: Orthoptera)
  * Praying Mantids (Order: Mantodea)
  * Mayflies (Order: Ephemeroptera)
  * Stick insects and Leaf insects (Order: Phasmatodea)
  * Stoneflies (Order: Plecoptera)
  * Termites (Previously Order: Isoptera but now part of Order: Blattodea)
  * Thrips (Order: Thysanoptera)
  * True Bugs (Order: Hemiptera)
  * Web-spinners (Order: Embioptera)
  * Zorapterans (Order: Zoraptera)
  * [Division] Endopterygota
  * Bees, Wasps and Ants (Order: Hymenoptera)
  * Beetles (Order: Coleoptera)
  * Butterflies and Moths (Order: Lepidoptera)
  * Caddisflies (Order: Trichoptera)
  * Fleas (Order: Siphonaptera)
  * Flies (Order: Diptera)
  * Lacewings, Antlions, Snakeflies, Dobsonflies & Alderflies (Order: Neuroptera)
  * Scorpionflies (Order: Mecoptera)
  * Strepsipterans (Order: Strepsiptera)

Non-insect Hexapods

There are three classes of invertebrate known as non-insect hexapods:-

- Proturans (Class: Protura)
- Springtails (Class: Collembola)
- Two-pronged bristletails (Class: Diplura).

Insect Classification

Fig. : An adult male Small White butterfly (Pieris rapae)*, just one of two species of butterfly that are called Cabbage Whites.*

People recognise things by giving them a name. This is also true for living things. Common ones have names from everyday language. For example, the Cabbage White Butterfly, or the hoverfly, or the dragonfly, or the ladybird. These names are fine for general everyday use, but they do have problems:-

1. There are two different types of butterfly that are called the Cabbage White.
2. There are over 250 different types of hoverfly in Britain.
3. The name dragonfly refers to a number of different species, not just to one.
4. In other parts of the World, ladybirds are not called ladybirds. For example, in North America ladybirds are called ladybugs.

There are other problems as well. One of these is that the name can be different in different parts of the country. For example, a woodlouse is called a *Jiggy Pig* in some parts of the country, and by other names elsewhere. Another problem is that animals that are rarely or never seen won't have an everyday name. Several hundred years ago, as science was developing, scientists started to name and describe living things in Latin. Often the process became tedious and at times the descriptions became the name! The situation became very chaotic. Eventually, a Swedish scientist called Carl von Linné tried to sort it all out. (The Latin version of his name is Carolus Linnaeus-he lived 1707-1778.) He devised a system using

2 names. One of these is the species name (in older writings it may be called the trivial name). This species name is more or less unique to that specific type of plant or animal. Also, he wanted his system to show how different things were related. He therefore grouped different species that he thought were related into a genus. Hence, each different type was called a species and had two names:-

Genus Species

The genus name is written first and begins with a capital letter. The specific (or species) name is written second. It does not have a capital letter. Sometimes the species is named after someone. In these cases the name may begin with a capital letter. The words used for the names were generally in Latin, the language of science and learning at the time. To make it stand out, the name was underlined when written. When typesetting a book, the printer will put underlined words into *italics*. Hence, in a book, the scientific names are in *italics*.

The Latin for 2 is "bi" (e.g. a bicycle has 2 wheels). A Latin term for name is "nome". If these two words are put together, we have the word "binomial". Hence, the two-name system devised by von Linné became known as the "Binomial System".

He tried to show how different genera were related by grouping them into a family. A number of different families were grouped into an order, orders into class, classes into a phylum, and several phyla into kingdoms. The two main kingdoms are the plants and the animals This list of groups (or hierarchy) may be shown as follows:-

- Species-Many characteristics the same.
- Genus-The number of shared features (or characteristics) gets less as we go down the list.
- Family
- Order
- Class
- Phylum
- Kingdom-Few characteristics the same.

This system tries to show how the different types of living things are related, it is called a "Natural System of Classification". He published his ideas in his book "*Systema Naturae*" in 1735. This was revised a number of times. The tenth edition, published in 1758, is the most important. Modern classification in Zoology is based on this.

Linnaeus was influenced a lot by the earlier writings of the English naturalist John Ray (1627-1705). In fact, much of Linnaeus' work is based on, and is an extension of, Ray's works. This is particularly true of the classification of plants. Linnaeus's ideas on plant classification were published in his work "Species

Plantarum", published in 1753. The Binomial name is followed by the name (or abbreviation) of the person who first described the species. This person is called the author.

The Binomial System is used by scientists all over the world. This means that the names are international. Therefore, if you write an article about *Pieris brassicae*, all scientists will know what species you are writing about. This would not be true if you wrote about the Cabbage White Butterfly.

Many people are "put off" by the scientific names. This is a shame because they are generally easy and many people use them without realising when they talk about plants. The best way is to use them from the beginning. It doesn't matter if you pronounce them wrongly, eventually you will learn the correct pronunciation. The important thing is to use them. You will probably have trouble at first. Do persevere as you will soon get the hang of them. Then you will wonder why you were wary of them in the first place!

Insects & Man

Insects as an opportunity: Insects pollinate 75% of all crop plant species. Here a bumblebee enters the flower of a Fox Glove. Insects have an important role as part of the biosphere-the living planet-and they make up four-fifths of all animal species. But what difference do they make to our lives?

Insects can be both a problem, and an opportunity!

Insects as a Problem

Many insect species compete with us for food. Many of these have been encouraged due to human activities, but then they become a problem. We plant crops which some insect species love, they thrive on them-and that's when we start to call them pests. Humans have caused some insect species to be moved from one part of the world to another-deliberately or by accident. The Large White butterfly is an example of a butterfly which was introduced to the USA and is now a pest on cabbages there. Another example is the European Corn Borer, *Ostrinia nubilalis*, a Pyralid moth which is an important pest of corn in the Midwest of the USA. The species was introduced there from Europe in the early 20th century and now costs US farmers over a billion dollars a year.

If the introduced insect species has no natural enemies in their new home, they can become a real problem. A recent UK example is the Harlequin ladybird (*Harmonia axyridis*), which has found its way to Britain and eats other ladybirds. In other countries where it has been deliberately introduced to control insect pests it has become a real problem, threatening to wipe out many native insects.

Some insects can sting, bite, or transmit disease to humans. Sometimes this is because they find us tasty-mosquitoes for example gain nourishment by sucking blood, and in the process they can transmit diseases, such as malaria-which is caused by the protozoan *Plasmodium falciparum* being transmitted in the saliva of the Anopheles mosquito. The flea *Chenopsylla cheopsis* had a profound effect on human history when it travelled from Africa in the ships of the middle ages, spreading the Black Death (also known as the Bubonic Plague) across Europe. It took four hundred years for the European population to recover from the effects of the Black Death. The illness is now known to be caused by the bacterium *Yersinia pestis*, which was carried by the flea, which in turn was carried by the black rat on ships. So, when it comes to insects, it's important to understand the consequences of human actions.

Many pests-such as the Colorado Beetle or the Gypsy Moth-can be controlled by spraying with chemical pesticides, which kill most insect species on the crop. But-would you rather eat perfect food sprayed with chemicals, or food where the pests are controlled by more natural means?

Insects as an Opportunity

Insects are very important as pollinators. Bees are perhaps the best example of this. Honey Bees aren't bred in captivity just for the honey they produce-they are vital to pollinate crops. Without them, many crops would fail, as pollination is needed before many crops and fruits that we eat can be formed.

Other examples of useful insects are the Silkworm Moth, *Bombyx mori*, from which silk is obtained. This species has been reared by humans for such a long time that it has become domesticated-it would be unlikely to survive in the wild if released today. Insects are also useful as models in scientific research. The Fruit Fly *Drosophila melanogaster* is a good example. It breeds rapidly, and can produce very many generations per year in the laboratory, so it is ideal for the study of genetics and evolution. In fact, many genetic discoveries of great importance to medicine have been made using this little fruit fly.

Insects can be useful to the gardener too-ladybirds, for example, are the gardener's friend because they eat aphids. Hoverflies and wasps also prey on other insects in the garden. These species are called predators. Others that are natural enemies to garden pests include parasitoids, which lay eggs in the host insect so as to provide food for its developing young. Many kinds of insects are also food for songbirds and other birds and animals. When there aren't enough insects around, the result is fewer birds and mammals. And insects can also be eaten by man! This is called entomophagia.

Become an Entomologist

Fig. : Photograph of the poisonous grasshopper, Dictyophorus spumans.

An entomologist is someone who takes an interest in insects, and studies them. They may be any age, and do not need to be professionals. Indeed, it is a notable feature of this subject that some of the greatest entomologists have been amateurs. Like any form of natural history, being an entomologist means taking an interest in the World around you.

You don't need any expensive equipment to start finding out more about insects-you just need to get outside and explore. Television documentaries showing lions hunting gazelles are impressive but, in your back garden or local park, you might find lacewing larvae devouring aphids and attaching the bodies of its victims to its back or solitary bees digging an underground nest and collecting pollen for its young. All you have to do is look!

Why are Entomologists Important?

It has been estimated that there are seven mammologists for every known species of mammal. However, there is only one entomologist for every 425 described species of insect. Entomologists have so far described one million species of insect but conservative estimates suggest that there could be another four to six million species waiting to be described.

Lepidoptera

Lepidoptera is an order of insects that includes moths and butterflies. It is one of the most speciose orders in the class Insecta, encompassing moths and the three superfamilies of butterflies, skipper butterflies, and moth-butterflies. Members of the order are referred to as lepidopterans. A person who collects or studies this order is referred to as a lepidopterist. This order has more than 180,000 species in 128 families and 47 superfamilies. The name is derived from Ancient Greek scale and wing. Estimates of species suggest that the order may have more species and is among the four largest, successful orders, along with the Hymenoptera, Diptera, and the Coleoptera.

General Characteristics

Lepidopterans like all holometabola, undergo complete metamorphosis, going through a four-stage life cycle: egg, larva/caterpillar, pupa/chrysalis, and imago/adult. Their lifecycle can include a inactive periods or diapauses in any of the pre-adult stages that help them overcome unsuitable environmental conditions. A species of Lymantrid moth, *Gynaephora groenlandica*, found in the Arctic has an exceptional life cycle that was thought to take 14 years but now estimated to take 7 years. The larvae, caterpillars, have a toughened (sclerotized) head capsule, chewing mouthparts, and a soft body, that may have hair-like or other projections, 3 pairs of true legs, and additional prolegs (up to 5 pairs). They can be confused with the larvae of sawflies. Lepidopteran larvae can be differentiated by the presence of crochets on the prolegs which are absent in the Symphyta (sawflies). Most caterpillars are herbivores, but a few are carnivores (some eat ants or other caterpillars) and detritivores. Adults have two pairs of membranous wings covered, usually completely, by minute scales. In some species, wings are reduced or absent (often in the female but not the male). Antennae are prominent. In moths, males frequently have more feathery antennae than females, for detecting the female pheromones at a distance. The Trichoptera (caddisflies) which are a sister group of the Lepidoptera have scales, but also possess caudal cerci on the abdomen, a feature absent in the Lepidoptera.

Adult mouth parts prominently include the proboscis formed from maxillary galeae and are adapted for sucking nectar. Some species have reduced mouth parts (some species do not feed as adults), and others have them modified to pierce and suck blood or fruit juices (some Noctuids). Mandibles are absent in all except the Zeugaloptera which have chewing mouthparts. The maxillary palpi are reduced and consist of up to five segments. They are conspicuous in some of the more primitive families and are often folded. The labial palpi are more prominent and upward pointed.

The three thoracic segments are fused and consist of non-movable sclerites. The wings arise from the meso-and meta-thoracic segments and are similar in size in the primitive groups. In the more recent groups, the meso-thoracic wings are larger with more powerful musculature at their bases and more rigid vein structures on the costal edge. In the Noctuoidea, the metathorax is modified with a pair of tympanal organs. There are a variety of wing coupling mechanisms that connect the forewings and the hindwings. The more primitive groups have an enlarged lobe, jugum, at the base of the forewing that folds under the hindwing in flight. Other groups have a frenulum on the hindwing that hooks under a retinaculum on the forewing. In some groups such as the Psychidae, Lymantriidae, the females are flightless and have reduced wings.

The abdominal segments 7-10 or 8-10 are modified to form the external genitalia. The abdomen is connected to the thorax in the more recent families by muscles connectd to projections from the abdominal sternite 2. Paired hearing organs at the base of the abdomen occur in the Pyraloidea and Geometroidea. Males have glandular organs such as expandable hair brushes or tufts, or as thin-walled, eversible sacs (coremata), from the intersegmental membranes. The genitalia are complex and provide the basis for species discrimination in most families and also in family identification. Primitive groups have a single genital aperture near the end of the abdomen through which both copulation and egg laying occur. This character is used to designate the Monotrysia. Hepialidae and related families have an external groove that carries sperm from the copulatory opening (gonopore) to the (ovipore) and are termed exoporian. The remaining groups have an internal duct that carry sperm and form the Ditrysia, with two distinct openings each for copulation and egg-laying. It also has antennaes.

Families

There are about 69 families in this order with variations depending on the taxonomic treatment. The Lepidoptera are divided into several suborders, the largest being Glossata, the vast majority of which are Ditrysia.

Several other classifications of lepidopteran families are used in older literature. These include the Rhopalocera (club-horned) consisting of what are commonly called butterflies and the Heterocera (varied-horned) consisting of the moths. However, Rhophalocera is a natural (monophyletic) group, while Heterocera is a paraphyletic assemblage. Another non-standard classification separates the Lepidoptera into Microlepidoptera for the smaller species (mostly moths) and Macrolepidoptera for the larger species.

History of Study

Linnaeus in *Systema Naturae* (1758) recognized three divisions of the Lepidoptera: *Papilio, Sphinx,* and *Phalaena* with seven subgroups in *Phalaena*. These persist today as 9 of the superfamilies of Lepidoptera. Other works on classification followed including those by Denis & Ignaz Schiffermüller (1775), Fabricius (1775) and Pierre André Latreille (1796). Jacob Hübner described many genera, and the Lepidopteran genera were catalogued by Ochsenheimer and Treitschke in a series of volumes on the Lepidopteran fauna of Europe published between 1807 and 1835. G.A.W. Herrich-Schaffer, and Edward Meyrick (1895) based their classifications primarily on wing venation.

Among the first entomologists to study fossil insects and their evolution was Samuel Hubbard Scudder (1837-1911), who worked on butterflies. He published a study of the Florissant deposits of Colorado. Andreas V. Martynov (1879-1938) recognized the close relationship between Lepidoptera and Trichoptera in his studies on phylogeny.

Lepidoptera tend not to be as common as some other insects in the habitats that are most conducive to fossilization, such as lakes and ponds, and their juvenile stage has only the head capsule as a hard part that might be preserved. Yet there are fossils, some preserved in amber and some in very fine sediments.

Leaf mines are also seen in fossil leaves, although the interpretation of them is tricky. The earliest fossil is *Archaeolepis mane* from the Jurassic, about 190 million years ago in Dorset, UK. It consists of wings and shows scales with parallel grooves under a scanning electron microscope and the characteristic wing venation pattern shared with Trichoptera. Only 2 more sets of Jurassic Lepidopteran fossils have been found, and 13 sets in the Cretaceous. From there, many more fossils are found from the Tertiary, and particularly the Eocene Baltic amber.

Major contributions in the 20th century included the creation of the monotrysia and ditrysia (based on female genital structure) by Borner in 1925 and 1939. Willi Hennig (1913-1976) developed the cladistic methodology and applied it to insect phylogeny. Niels P. Kristensen, E. S. Nielsen and D. R. Davis studied the relationships among monotrysian families and Kristensen worked more generally on insect phylogeny and higher Lepidoptera too..

While it is often found that DNA-based phylogenies differ from those based on morphology, this has not been the case for the Lepidoptera; DNA phylogenies correspond to a large extent to morphology-based phylogenies. Many attempts have been made to group the superfamilies of the Lepidoptera into natural groups, most of which fail because one of the two groups is not monophyletic: Microlepidotera and Macrolepidoptera, Heterocera and Rhopalocera, Jugatae and Frenatae, Monotrysia and Ditrysia.

Phylogeny

It has long been noted that the Lepidoptera and the Trichoptera (caddisflies) share many similarities that are lacking in other insect orders. Among these are:

- females, rather than males, are heterogametic (i.e. their sex chromosomes differ)
- dense setae on the wings (modified into scales in Lepidoptera)
- a particular wing venation pattern on the forewings
- larvae with mouth structures and glands to make and manipulate silk.

Thus the two sister orders are grouped into the Amphiesmenoptera. The group probably evolved in the Jurassic, diverging from the extinct Necrotaulidae. Lepidoptera differ from the Trichoptera in several features, including wing venation, form of the scales on the wings, loss of the cerci, loss of an ocellus, and changes to the legs. The oldest, most basal lineages of Lepidoptera have as adults, not the curled tongue or proboscis characteristic of most members of the order, but chewing mandibles. Micropterigidae larvae feed on decaying leaves (much like the Trichoptera), fungi, liverworts or live leaves. The adults chew pollen or spores of ferns. In the Agathiphagidae, larvae feed inside seeds of kauri pines, and in Heterobathmiidae the larvae mine leaves of *Nothofagus*, the southern beech. These families also have mandibles in the pupal stage, which help the pupa emerge from the seed or cocoon just before adult emergence.

The Eriocraniidae have a short coiled proboscis in the adult stage, and retain mandibles for the purpose of escaping the cocoon, but they are non-functional thereafter. They, and most of the other non-ditrysian families, are primarily leaf miners in the larval stage. In addition to the proboscis, there is a change in the scales among these basal lineages, with later lineages showing more complex perforated scales.

With the evolution of the Ditrysia in the mid-Cretaceous, there was a major reproductive change. The Ditrysia, which comprise 98% of the Lepidoptera, have two separate openings for reproduction in the females (as well as a third opening for excretion), one for mating, and one for laying eggs. The two are linked internally by a seminal duct. (In more basal lineages there is one cloaca, or later, two openings and an external sperm canal.) Of the early lineages of Ditrysia, Gracillarioidea and Gelechioidea are mostly leaf miners, but more recent lineages feed externally. In the Tineoidea, most species feed on plant and animal detritus and fungi, and build shelters in the larval stage.

The Yponomeutoidea is the first group to have significant numbers of species whose larvae feed on herbaceous plants, as opposed to woody plants. They evolved about the time that flowering plants underwent an expansive adaptive radiation in the mid-Cretaceous, and the Gelechioidea that evolved at this time also have great diversity. Whether the processes involved co-evolution or sequential evolution, the diversity of the Lepidoptera and the angiosperms increased together.

In the so-called "macrolepidoptera", which constitutes about 60% of Lepidopteran species, there was a general increase in size, better flying ability (via changes in wing shape and linkage of the forewings and hindwings), reduction in the adult mandibles, and a change in the arrangement of the crochets (hooks) on the larval prolegs, perhaps to improve the grip on the host plant. Many also have tympanal organs, that allow them to hear.

These organs evolved eight times, at least, because they occur on different body parts and have structural differences. The main lineages in the macrolepidoptera are the Noctuoidea, Bombycoidea, Lasiocampidae, Mimallonoidea, Geometroidea and Rhopalocera. Bombycoidea plus Lasiocampidae plus Mimallonoidea may be a monophyletic group. The Rhopalocera, comprising the Papilionoidea (Butterflies), Hesperioidea (skippers), and the Hedyloidea (moth-butterflies), are the most recently evolved. There is quite a good fossil record for this group, with the oldest skipper about 56 million years old.

Butterflies and Moths (Order Lepidoptera)

A little known passion of Sir Winston Churchill's was his concern for the diminishing numbers of British butterflies. In 1946 he planned a butterfly garden to increase the numbers of common species around his house at Chartwell, Kent. About 1,500 chrysalids were hatched each year in a summer house. Churchill would spend hours waiting for the moment when the butterflies emerged. Once on the wing he set them free. The Order Lepidoptera comprises over 16,000 species of butterflies and moths, with most of these being moths. Only the Coleoptera (Beetles) form a larger order of any animal.

The Lepidoptera are probably the most widely studied order of invertebrates, and have been for more than 400 years. More books have been written about them than on any other group of invertebrates. Indeed, the long history of their study by Lepidopterists has itself been the subject of a few books, and is a fascinating study on its own! Scientifically, there is no real distinction between butterflies and moths. However, in general, butterflies are day flying, whilst moths fly mainly at night. Butterflies are usually slimmer bodied, and have thin antennae with distinct clubs at the end. Moths have antennae of various designs, from thin and tapering to wide and 'feathery'. Feathery antennae are found in male moths, and help them to sniff out females!

Due to their often bright colours and association with warm sunny days, butterflies have tended to catch the popular imagination over the centuries, more than any other insects. They can even be found adorning some ancient Egyptian tombs.

Moths are not always thought of so highly, no doubt due to their nocturnal habits and duller colours. However, many moths are brightly coloured and fly during the day. On the other hand, some butterflies are active at twilight, and some are no more brightly coloured than many moths. Even the tiniest moths can look spectacularly beautiful when viewed closely.

Moths are often arbitrarily split into two groups-the larger moths, or *macrolepidoptera* (macros) and the smaller moths, or *microlepidoptera* (micros). While the micros tend to be more primitive in evolutionary terms, this is not always the case; and, some micros are in fact larger than some of the macros! So, like the division between moths and butterflies, this distinction is arbitrary too, and has no scientific basis.

The one advantage these subdivisions do have is that people can study and get to know the Lepidoptera in stages! Often, people start with the butterflies (around 70 species in the UK) then their curiosity takes them on to the larger moths (around 900 species) and finally to the microlepidoptera (around 1,500 species). Most of these species are resident in the UK, others are recorded when they migrate here every year from warmer climes.

Main Characters of Order

- The scientific name of the order, Lepidoptera, is derived from one of their main characteristics, namely their having wings covered in tiny scales (from the Greek lepidos = scale and pteron = wing). Indeed, it is these coloured scales which give them their patterns. These scales are specially modified flattened hairs.
- The Lepidoptera undergo complete metamorphosis, i.e. ova (egg), from which emerge larvae (caterpillars), which become the quiescent pupae (chrysalis) from which emerge the imago (winged adult). This lifecycle can take anywhere between a few weeks to more than a year, depending upon the species.
- Lepidoptera are 'typical' insects, in that they have 4 wings, 6 legs, 2 antennae and a body divided into 3 sections-a head, thorax and abdomen. The leg and wings are attached to the thorax. In a few species of moths, the females have evolved to become wingless.
- Most butterflies and moths feed through a specialised tube formed by some of the mouthparts, known as a proboscis. Nectar is the usual food for adults.
- Sense organs on the feet can taste certain food substances with a greater sensitivity than the human tongue.

- The wings consist of an upper and lower membrane supported by a system of hollow veins.
- Most Lepidoptera larvae feed exclusively on plant matter, but a few are carnivorous for at least part of their life. Some species feed on a wide variety of plants, whilst others are willing to accept only one or two.
- The larvae moult several times, usually 4, 5 or 6 depending upon the species.
- The final moult reveals the pupa, which can be attached to part of the foodplant or other nearby item, unattached amongst debris such as leaf litter, or in a silk lined chamber underground.
- Survival strategies of butterflies and their earlier stages include camouflage, toxic defence such as being distasteful/harmful or mimicking species that are distasteful/harmful.
- Some species are quite small, having a wingspan of only a few millimetres, whilst others are giants among the insect world, being 30cm across.

7

The Forensic Entomology

Forensic entomology contains three aspects: medicocriminal entomology, urban entomology, and stored product entomology. This article focuses more on the medicocriminal aspect and how DNA is analysed with various blood feeding insects.

Blood Meal Extraction

To extract a blood meal from the abdomen of an insect to isolate and analyse DNA, the insect must first be killed by placing it in 96% ethanol. The killed insect can be stored at -20°C until analysis. When it is time for analysis, the DNA must then be extracted by dissecting the posterior end of the abdomen and collecting 25mg of tissue. The cut in the abdomen should be made with a razor blade as close to the posterior as possible to avoid the stomach. Using a DNA extraction kit, the DNA is extracted from the tissue. If the DNA is mixed with samples from more than one individual, it is separated using a species specific primer. Once extracted and isolated, the DNA sample goes through a polymerase chain reaction (PCR), is amplified and identified. PCR works by analysing species specific mitochondrial DNA. PCR is currently the most commonly used method of species identification. This results from the fact that it is very sensitive in that it requires only a small amount of biological material, and can also utilize material that is not particularly fresh.

The sample can be frozen and stored while still remaining usable for later PCR. DNA requires one hour to reach the abdomen of an insect, so DNA can be amplified one to forty-four hours after an insect feeds. Some research suggests that the source of a blood meal can be determined up to two months post feeding. To amplify DNA, it must first be denatured by exposing it to a 95°C temperature for one minute, followed by thirty cycles of thirty-second 95°C exposures. Then denatured DNA is mixed with a specific primer. A chromatograph is conducted on 2% agarose gel, stained, and viewed with UV fluorescence. The DNA is identified by looking for genome specific repetitive elements and by comparing it with known examples.

Haematophagous Insects of Forensic Importance

Humans are constantly fed on by haematophagous insects. The ingested blood can be recovered and used to identify the person from which it was taken. Bite marks and reactions to bites can be used to place a person in an area where those insects are found.

Order Diptera

The following among the flies (Diptera) have been utilized:

- Mosquitoes, Family Culicidae : Due to erratic feeding habits, mosquitoes could potentially provide DNA evidence to many people in one area at a certain time. PCR analysis shows that identification of bitten individuals is possible with a low error rate, although multiple mosquitoes would be needed. The insects would need to be collected as soon as possible due to the insect's high mobility and constant feeding. Research is centred on the mosquito due its widespread presence and affinity for feeding on humans.
- Biting midges, Family Ceratopogonidae
- Tsetse flies, Family Glossinidae
- Sheep keds, Family Hippoboscidae
- Stable and horn flies, Family Muscidae
- Sand flies, Family Psychodidae, Subfamily Phlebotominae
- Snipe flies, Family Rhagionidae
- Black flies, Family Simuliidae
- Horse flies, Family Tabanidae.

Order Siphonaptera

Listed here are fleas commonly encountered by humans that could potentially be used for DNA identification.

- Sticktight and chigoe flea, Family Hectopsyllidae (formerly Tungidae)
- Cat flea (*Ctenocephalides felis*)
- Northern rat flea (*Nosopsyllus fasciatus*)
- Human flea (*Pulex irritans*)
- Oriental rat flea (*Xenopsylla cheopis*).

Order Hemiptera

- Bedbug (*Cimex lectularius*): *Cimex lectularius* is an obligate parasite of humans. Testing a sample of a residence's bed bug population and screening for bites could reveal possible recent visitors to the structure, as they have been observed to feed approximately once a week in temperate conditions.

A recent re-emergence of bedbug populations in North America as well as growing interest in the field of forensics may prove bedbugs to be useful investigative tools. Recent studies have revealed that human DNA can be recovered from bed bugs for up to 60 after feeding, thus demonstrating the potential use of this insect in forensic entomology

- Assassin bugs, Family Reduviidae

Order Phthiraptera

Lice can be indicators of contact with another person. Many species closely associated with humans can be easily transferred between individuals. DNA identification of multiple individuals using blood meals from body and head lice has been demonstrated in laboratory settings.

Suborder Anoplura

- Head louse (*Pediculus humanus capitis*)
- Body louse (*Pediculus humanus humanus*)
- Pubic louse (*Phthirus pubis*).

Other Arthropods

Order Ixodida

Due to the low probability of a tick detaching and falling to the ground at the scene of the crime, these may not be highly useful regardless of the large amount of blood and lymph they ingest. However, should an engorged tick be found in an area of interest, it would likely contain sufficient genetic material for identification.

Analysis of Collected DNA

DNA identification of species can be a very useful tool in forensic entomology. Although it does not replace conventional identification of species through visual identification, it can be used to differentiate between two species of very similar or identical physical and behavioural characteristics. A thorough identification of the species through conventional methods is needed before an attempt at DNA analysis. This DNA can be obtained from practically any part of the insect, including the body, leg, setae, antennae, etc. There are about one million species described in the world and many more that have still not been identified. A project termed "the barcode of life" was launched by Dr. Paul Herbert, where he identified a gene that is used in cell respiration by all species, but is different in every species. This difference in sequence can help entomologists easily identify two similar species.

DNA sequencing is basically done in three steps: polymerase chain reaction (PCR), followed by a sequencing reaction, then gel electrophoresis. PCR is a step that cleaves the long chain of chromosomes into much shorter and workable pieces.

These pieces are used as patterns to create a set of fragments. These fragments are different in length from each other by one base which is helpful in identification. Those sets of fragments are then separated by gel electrophoresis. This process uses electricity to separate DNA fragments by size as they move through a gel matrix. With the presence of an electric current the negative DNA strand marches toward the positive pole of the current. The smaller DNA fragments move through the gel pores much more easily/faster than larger molecules. At the bottom of the gel the fragments go through a laser beam that emits a distinct colour according to the base that passes through.

Case Studies

Case Study #1

Research has already demonstrated the link between forensic entomology and society by amplifying human DNA from blood meals through various methods. A case in Italy involving the murder of a woman used this technique to identify a suspect.

The woman's body was discovered partially covered by sand on a beach in Sicily. Investigators suspected a prominent businessman, whose car was seen in the area the night of the murder. A search of the suspect's home produced no substantial evidence, apart from a mosquito blood meal stain on the wall. Technicians absorbed the stain onto wet filter paper and then scratched the rest of the blood material off the wall. They also collected the insect's remains in a tube for species identification. They extracted DNA from the blood sample and performed PCR. They identified the DNA as the victim's, which placed the victim in the vicinity, if not in, the suspect's home. Experts identified the mosquito as *C. pipiens*, a species not known to travel far distances, such as between the suspect's home and the crime scene. This evidence, along with grains of sand and leaf fragments on the suspect's clothing that matched samples from the beach, helped convict the suspect of second degree murder. This also relates forensic entomology and the law.

Case Study #2

Forensic Entomologist Dr. Jeff Wells from West Virginia University worked on this specific case.

The evidence (no details were provided as to the circumstances surrounding their collection) was received as two shipping forms and 17 gelatin capsules, each of which contained a hand-written label and either a dried fly larva (maggot), what appeared to be a fragment of adult insect cuticle, or what appeared to be a fragment of insect pupal cuticle (puparium). A DNA extraction procedure was performed on each specimen using the commercial kit sold by Qiagen. Each sample was subjected to chemical digestion but no deliberate mechanical damage. The middle portion was cut out of the body of the larvae specimens for digestion and DNA extraction following the manufacturer's instructions. Any tissue from the middle section that remained following digestion, plus the unprocessed anterior and posterior sections of each larva have been stored with the original label in a screw-cap vial filled with 95% ethanol.

All material from a non-larval specimen was digested. Any fragments that remained following the extraction procedure were stored in ethanol as above. The outside of each vial was labelled with the extraction number. Mitochondrial DNA sequence analysis was attempted on each DNA extract as described in Wells and Sperling using PCR and sequencing primers C1-J-1751 and C1-N-2191.

The results from the PCR experiment showed that the DNA analysis was successful with eight larval specimens. Analysis on the insect cuticle proved unsuccessful as no mitochondrial DNA was able to be extracted. The portion of the mitochondrial DNA molecule characterized by these procedures corresponds to base positions 1752–2190 of the commonly used system first established for *Drosophila yakuba*. The sequences (also known as haplotypes) obtained ranged from 414 bases to 439 bases in length. Below is an example of a larva specimen DNA analysis.

Specimen 4

Species determination was done based on a mitochondrial DNA (mtDNA) haplotype. Three specimens produced the same haplotype, matching the previously published haplotype for the calliphorid fly species *Cynomya cadaverina* and therefore being identified as such. Several specimens produced three very closely related haplotypes, indicating that they are all the same species. However, these haplotypes are not closely related to any previously published or unpublished data. Therefore these specimens cannot be identified based solely on haplotype data, although they are definitely not members of the forensically important and extensively studied fly families Calliphoridae or Sarcophagidae. There are several common carrion-feeding species in the family Muscidae that have not been investigated using DNA methods, and one of these species, *Hydrotea dentipes*, fits the general appearance of these specimens. Given the current state of the science these specimens cannot be identified using DNA analysis.

Conclusion

Forensic entomology is a very important aspect for law enforcement. With the magnitude of information that can be gathered, investigators can more accurately determine time of death, location, how long a body has been in a specific area, if it has been moved, and other important factors. Because of this range of uses, forensic entomology is an extremely significant tool used by law enforcement officials to aid in numerous cases. As this branch of entomology progresses it will become a key facet in all investigations due to its steadily rising popularity and usefulness.

Respiration

Tracheae & Oxygen Diffusion

Insects, in general, do not have an oxygen-carrying chemical in their blood so oxygen reaches cells by other means.

Most insects have a waterproof cuticle but some insects live in moist areas and are sedentary. Their cuticle is permeable to water and they obtain sufficient oxygen by diffusion across their cuticle.

However, most insects have a special respiratory system comprising a system of internal tubes, called trachea, which branch and rebranch. Very fine branches, tracheoles, penetrate individual cells.

Trachae have spiral stiffening-like vacuum cleaner hose-to prevent collapse.

Air enters from the outside through a series of openings, spiracles. Typically there are 2 pair of spiracles laterally on the thorax and 8 pair laterally on the abdomen.

O_2 from spiracles —> tracheae —> tracheoles —> cells

Rate of gas diffusion is determined by

- Differences in concentration at each end of the system
- Molecular weight of the gas. (The lower the molecular weight, the faster the diffusion rate)
- Permeability of the substrate:
  - * O_2 in air = 1.1 ml/min/cm^2/atm
  - * O_2 in tissue = 1.4 x 10 $^{-5}$ ml/min/cm^2/atm
  - * i.e. Diffusion through air is more than 100,000 x the rate of diffusion through tissue.

So most of the path for oxygen to diffuse from the spiracle to the cell is through the air-filled tracheal tubes.

Spiracles

Insects need to avoid water loss through their spiracles and also to prevent contamination by dust etc.

Spiracles are therefore usually equipped with opening and closing devices and filtering lobes or hairs in an atrium before the beginning of the trachae.

Very active insects have internal air sacs, as extra reservoirs, as part of their tracheal system. They also may employ mechanical ventilation along the larger trachae. Bees and wasps may extend and telescope their abdomens to pump air along.

Aquatic Respiration

Many aquatic insects must return periodically to the surface to breathe.

Mosquito larvae have tubes connecting to the air at the water surface. There are hydrofuge hairs (water resistant) associated with the spiracles, to break the surface tension of the water and to keep water out when the insect is underwater.

Other insects, such as stone fly, dragonfly and mayfly nymphs have tracheal gills whereby oxygen diffuses from the water to fine, surface trachae.

Water beetles take a bubble of air with them underwater. Their spiracles open into the bubble As the oxygen in the bubble is used up, the oxygen pressure in the bubble drops to that of the surrounding water so that more oxygen diffuses into the bubble from the water. In this way the bubble remains usable for a long time. Bloodworms, (midge larvae) live in the mud of stagnant pools where there is very little oxygen. These insect larvae are one of the few which have haemoglobin in their blood and this haemoglobin is more efficient than ours.

Small internal parasites may diffuse oxygen across their cuticle. Others may have spiracles at the end of siphons which open into the trachae of their host or penetrate the skin of their host into the air.

Research Results For 'Respiration'

Allantois

The allantois is a structure appearing during the early development of vertebrate animals-Reptiles, Birds, and Mammalia

It is largely made up of blood-vessels, and, especially in birds, attains a large size. It forms the inner lining to the shell, and may thus be viewed as the surface by means of which the respiration of the embryo is carried on. In Mammalia the allantois is not so largely developed as in birds, and it enters largely into the formation of the placenta.

Animal

An animal is an organized and sentient living being. Life in the earlier periods of natural history was attributed almost exclusively to animals. With the progress of science, however, it was extended to plants. In the case of the higher animals and plants there is no difficulty in assigning the individual to one of the two great kingdoms of organic

nature, but in their lowest manifestations, the vegetable and animal kingdoms are brought into such immediate contact that it becomes almost impossible to assign them precise limits, and to say with certainty where the one begins and the other ends. From form no absolute distinction can be fixed between animals and plants.

Many animals, such as the sea-shrubs, sea-mats, etc, so resemble plants in external appearance that they were looked upon as such. With regard to internal structure no line of demarcation can be laid down, all plants and animals being, in this respect, fundamentally similar; that is, alike composed of molecular, cellular, and fibrous tissues. Neither are the chemical characters of animal and vegetable substances more distinct.

Animals contain in their tissues and fluids a larger proportion of nitrogen than plants, whilst plants are richer in carbonaceous compounds than the former. In some animals, moreover, substances almost exclusively confined to plants are found. Thus the outer wall of Sea-squirts contains cellulose, a substance largely found in plant-tissues; whilst chlorophyll, the colouring-matter of plants, occurs in Hydra and many other lower animals.

Power of motion, again, though broadly distinctive of animals, cannot be said to be absolutely characteristic of them. Thus many animals, as oysters, sponges, corals, etc, in their mature condition are rooted or fixed, while the embryos of many plants, together with numerous fully developed forms, are endowed with locomotive power by means of vibratile, hair-like processes called cilia. The distinctive points between animals and plants which are most to be relied on are those derived from the nature and mode of assimilation of the food. Plants feed on inorganic matters, consisting of water, ammonia, carbonic acid, and mineral matters. They can only take in food which is presented to them in a liquid or gaseous state.

The exceptions to these rules are found chiefly in the case of plants which live parasitically on other plants or on animals, in which cases the plant may be said to feed on organic matters, represented by the juices of their hosts. Animals, on the contrary, require organized matters for food. They feed either upon plants or upon other animals. But even carnivorous animals can be shown to be dependent upon plants for subsistence; since the animals upon which Carnivora prey are in their turn supported by plants. Animals, further, can subsist on solid food in addition to liquids and gases; but many animals (such as the Tapeworms) live by the mere imbibition of fluids which are absorbed by their tissues, such forms possessing no distinct digestive system.

Animals require a due supply of oxygen gas for their sustenance, this gas being used in respiration. Plants, on the contrary, require carbon dioxide. The animal exhales or gives out carbon dioxide as the part result of its tissue-waste, whilst the plant taking in this gas is enabled to decompose it into its constituent carbon and oxygen. The plant retains the former for the uses of its economy, and liberates the oxygen, which is thus restored to the atmosphere for the use of the animal. Animals receive their food into the interior of their bodies, and assimilation takes place in their internal surfaces. Plants, on the other hand, receive their food into their external surfaces, and assimilation is effected in the external parts, as are exemplified in the leaf-surfaces under the influence of sunlight. All animals possess a certain amount of heat or temperature which is necessary for the performance of vital action.

The only classes of animals in which a constantly-elevated temperature is kept up are birds and mammals. The bodily heat of the former varies from 100 degrees Fahrenheit to 112 degrees Fahrenheit and of the latter from 96 degrees to 104 degrees. The mean or average heat of the human body is about 99 degrees Fahrenheit, and it never falls much below this in health. Below birds animals are named cold-blooded, this term meaning in its strictly physiological sense that their temperature is usually that of the medium in which they live, and that it varies with that of the surrounding medium, Warm-blooded animals, on the contrary, do not exhibit such variations, but mostly retain their normal temperature in any atmosphere. The cause of the evolution of heat in the animal body is referred to the union (by a process resembling ordinary combustion) of the carbon and hydrogen of the system with the oxygen taken in from the air in the process of respiration.

Ascidia

Ascidia (named from the Greek, askos, a wine-skin) is a name given to the Sea-squirts or main section of the Tunicata, a class of animals of low grade, resembling a double-necked bottle, of a leathery or gristly nature, found at low-water mark on the sea-beach, and dredged from deep water attached to stones, shells, and fixed objects. One of the prominent openings admits the food and the water required in respiration, the other is the excretory aperture. A single ganglion represents the nervous system, placed between the two apertures. Male and female reproductive organs exist in each ascidian. They pass through peculiar phases of development, the young ascidian appearing like a tadpole-body. They may be single or simple, social or compound. In social ascidians the peduncles of a number of individuals are united into a common tubular stem, with a partial common circulation of blood. In these animals early evolutionists saw a link between the Mollusca and the Vertebrata.

Bee

The most important member of the family is the common hive or honey bee. It belongs to the warmer parts of the Eastern Hemisphere, but is now naturalized in the Western. A hive commonly consists of one mother or queen, from 600 to 800 males or drones, and from 15,000 to 20,000 working bees, formerly termed neuters, but now known to be imperfectly-developed females. The last-mentioned, the smallest, have twelve joints to their antennae, and six abdominal rings, and are provided with a sting; there is, on the outside of the hind-legs, a smooth hollow, edged with hairs, called the basket, in which the kneaded pollen or bee-bread, the food of the larvae, is stored for transit.

The queen has the same characteristics, but is of larger size, especially in the abdomen; she has also a sting. The males, or drones, differ from both the preceding by having thirteen joints to the antennae; a rounded head, with larger eyes, elongated and united at the summit; and no stings. According to Huber the working-bees are themselves divisible into two classes: one, the cirieres, devoted to the collection of provisions, etc; the other, smaller and more delicate, employed exclusively within the hive in rearing the young.

The mouth of the bee is adapted for both masticatory and suctorial purposes, the honey being conveyed thence to the anterior stomach or crop, communicating with a second stomach in which alone a digestive process can be traced. The queen, whose sole office is to propagate the species, has two large ovaries, consisting of a great number of small cavities, each containing sixteen or seventeen eggs. The inferior half-circles, except the first and last, on the abdomen of working-bees, have each on their inner surface two cavities, where the wax, secreted by the bee from its saccharine food, is formed in layers, and comes out from between the abdominal rings.

Respiration takes place by means of air-tubes which branch out to all parts of the body, the bee being exceedingly sensitive to an impure atmosphere. Of the organs of sense the most important are the antennae, deprivation of these resulting in a species of derangement. The majority of entomologists regard their function as in the first place auditory, but they are exceedingly ssensitive to tactual impressions, and are apparently the principal means of mutual communication.

Bees undergo perfect metamorphosis, the young appearing first as larvae, then changing to pupae, from which the imagosor perfect insects spring. Whether the offspring are to be female or male is said to be dependent upon the contact or absence of contact of the egg with the impregnating fluid received from the male and stored in a special sac communicating with the oviduct, unfertilized eggs producing males. The further question whether the offspring shall be queens or workers is resolved by the influence of environment upon function. The enlargement of a cell to the size of a royal chamber and the nourishment of its inmate with a special kind of food appear to be sufficient to transform an ordinary working-bee larva into a fully-developed female or queen-bee.

The season of fecundation occurs about the beginning of summer, and the laying begins immediately afterwards, and continues until autumn; in the spring as many as 12,000 eggs may be laid in twenty-four days. Those laid at the commencement of fine weather all belong to the working sort, and hatch at the end of four days. The larvae acquire their perfect state in about twelve days, and the cells are then immediately fitted up for the reception of new eggs. The eggs for producing males are laid two months later, and those for the females immediately afterwards. This succession of generations forms so many distinct communities, which, when increased beyond a certain degree, leave the parent hive to found a new colony elsewhere. Thus three or four swarms sometimes leave a hive in a season. A good swarm is said to weigh at least three kilograms. Besides the common bee there are the Apis fasciata, domesticated in Egypt, the Apis Ugustica, or Ligurian bee of Italy and Greece, introduced into England, etc.

The humble-bees, or bumble-bees, of which about forty species are found in Britain and over sixty in North America, belong to the genus Bombus, which is almost worldwide in its distribution. Of these species solitary females which have survived the winter commence constructing small nests when the weather begins to be warm enough; some of them going deep into the earth in dry banks, others preferring heaps of stone or gravel, and others choosing always some bed of dry moss. In the nest the bee collects a mass of pollen and in this lays some eggs. The cells in these nests are not the work of the old bee, but are formed by the young insects similarly to the cocoons of silk-worms; and when the perfect insect is released from them by the old bee, which gnaws off their tops, they are employed as honey-cups.

The humble-bees, however, do not store honey for the winter, those which survive until the cold weather leaving the nest and penetrating the earth, or taking up some other sheltered position, and remaining there until the spring. The first brood consists of workers, and successive broods are produced during the summer. The experiment of domesticating different kinds of wild bees has been tried with no satisfactory results. Some bees, from their manner of nesting, are known as 'mason bees,' 'carpenter bees,' and 'upholsterer bees.' Some of these bees cement particles of sand or gravel together with a viscid substance in forming their nests; others make burrows in wood. The leaf-cutter or upholsterer bee lines its burrow with bits of leaf cut out in regular shapes.

Branchiogasteropoda

The branchiogasteropoda are gasteropodous molluscs whose respiration is aquatic, being generally effected by means of external branchiae or gills. They include a great many animals with univalve shells, as whelks, limpets, cone-shells, periwinkles, cowries, etc, also sea-hares, sea-slugs, sea-lemons, and the heteropoda.

Carnivora

Carnivora is a term applicable to any creatures that feed on flesh or animal substances, but is now applied specially to an order of mammals which prey upon other animals. The head is small, the jaws powerful, and the skin is well covered with hair. Two sets of teeth, deciduous or milk and permanent, are always developed in succession, and in both sets incisors, canines, and molars are distinguishable. The stomach is simple and the alimentary canal short, thus making the body as light and slender as possible for the purpose of hunting and springing on its prey.

The muscular activity of the Carnivora is very great, their respiration and circulation very active, and their demand for food is consequently constant. Carnivora are often divided into Plantigrada, comprising the bears, badgers, raccoons, etc; Digitigrada, comprising lions, tigers, cats, dogs; and Pinnipedia or Pinnigrada, comprising the seals and walruses. The two former divisions are also classed together as Fissipedia. The typical Plantigrada are distinguished by their putting the whole sole of the foot to the ground in walking, while the Digitigrada walk on the tips of their toes. The Plantigrada are also less decidedly carnivorous, and feed much on roots, honey, and fruits. In the Pinnigrada the body is long and of a fish shape, the fore and hind limbs are short and form broad webbed swimming-paddles. The hind-feet are placed far back, and more or less tied down. to the tail by the integuments.

Cyclobranchiata

The Cyclobranchiata are an order of gasteropods, in which the branchiae or organs of respiration form a fringe around the body of the animal, between the edge of the body and the foot. The order consists principally of the limpets.

Entomology

Entomology is the branch of zoology dealing with insects. It was started as a science in 1705 by the publication of Ray's 'Methodus Insectorum'.

The name entomology comes from the Greek entoma, animals 'cut in', the transverse division or segmentation of the body being their most conspicuous feature.

The true insects are those animals of the division Arthropoda or Articulata distinguished from the other classes of the division by the fact that the three divisions of the body-the head, thorax, and abdomen-are always distinct from one another. There are never more than three pairs of legs in the perfect insect, and these are all borne upon the thorax. Each leg consists of from six to nine joints. The first of these is called the 'coxa,' and is succeeded by a short joint called the 'trochanter.' This is followed by a joint, often of large size, called the 'femur,' succeeded by the 'tibia,' and this has articulated to it the 'tarsus', which may be composed of from one to five joints.

Normally two pairs of wings are present, but one or other may be wanting. The wings are expansions of the sides of the second and third sections of the thorax, and are attached by slender tubes called 'nervures'.

In the beetles the anterior pair of wings becomes hardened so as to form protective cases for the posterior membranous wings, and are called in this condition ' elytra ' or ' wing-cases.' Respiration is effected by means of air-tubes or tracheas, which commence at the surface of the body by lateral apertures called 'stigmata' or 'spiracles,' and ramify through every part of the body. The head is composed of several segments amalgamated together, and carries a pair of feelers or 'antennae', a pair of eyes, usually compound, and the appendages of the mouth. The thorax is composed of three segments, also amalgamated, but generally pretty easily recognized. The abdominal segments are usually more or less freely movable upon one another, and never carry locomotive limbs; but the extremity is frequently furnished with appendages connected with generation, and which in some cases serve as offensive and defensive weapons (stings).

The organs of the mouth take collectively two typical forms, the masticatory and the suctorial, the former exemplified by the beetles, the latter by the butterflies, in which the mouth is purely for suction. The alimentary canal consists of the oesophagus or gullet, a crop, a gizzard, a stomach, and an intestine, terminating in a cloaca. There is no regular system of blood vessels; the most important organ of the circulation is a contractile vessel situated dorsally and called the 'dorsal vessel.' The nervous system is mainly composed of a series of ganglia placed along the ventral aspect of the body and connected by a set of double nerve-cords.

The sexes are in different individuals, and most insects are oviparous. Reproduction is generally sexual, but non-sexual reproduction also occurs. Generally the young are very different from the full-grown insect, and pass through a 'metamorphosis' before attaining the mature stage. When this metamorphosis is complete it exhibits three stages-that of the larva, caterpillar, or grub, that of the pupa or chrysalis, and that of the imago or perfect winged insect.

Insects have been divided into three sections-Ametabola,Hfemimetabola, and Holometabola, according as they undergo no metamorphosis, an incomplete one, or a complete one. The young of the Ametabola differ from the adult only in size. They are all destitute of wings; the eyes are simple and sometimes wanting. The Hemimetabola undergo an incomplete metamorphosis, the larva differing from the imago chiefly in the absence of wings and in size. The pupa is usually active, or if quiescent capable of movement. In the Holometabola the metamorphosis is complete, the larva, pupa, and imago differing greatly from one another in external appearance and habits. The larva is wormlike and the pupa quiescent.

The section Ametabola (which in the opinion of many naturalists are scarcely within the pale of the true Insecta) is divided into three orders-Anoplura (lice), Mallophaga (bird-lice), and Thysanura (springtails). The section Hemimetabola comprises the orders Hemiptera (cicadas, bugs, plant-lice, etc), Orthoptera , and Neuroptera. The Holometabola comprises the orders Aphaniptera (fleas), Diptera (gnats, bot-flies, gad-flies, mosquitos, house-flies, etc), Lepidoptera (butterflies and moths), Hymenoptera (bees, wasps, and ants), Strepsiptera (stylops, minute and parasites), and Coleoptera (ladybirds, glow-worms, cockchafers, weevils, and all of the beetle tribe).

A division is sometimes made into Mandibulate and Haustellate groups, the oral apparatus of the former being adapted for mastication, the latter for imbibition of liquid food. Both types are, however, sometimes modified, and occasionally combined.

Blood Vascular System

The blood and lymphatic vascular systems are classified as specialized connective tissue.

The main functions of the blood are to transport oxygen, nutrients and hormones to the tissues and to collect the waste products (carbon dioxide and waste metabolites) for removal from the body via the excretory system.

The cardiovascular system consists of the:

- Heart (muscular pump)
- Pulmonary circulation (system of blood vessels to and from the lungs)
- Systemic circulation (system of blood vessels bringing blood to and from all the other organs of the body).

Arteries are classified into two main groups:

* Conducting (Elastic Arteries).

These are large arteries closest to the heart (aorta, renal artery) with very high blood pressure and flow (320mm/sec in the aorta).

* Distributing (Muscular Arteries).

These are smaller diameter arteries with a slower blood flow. The arteries lead to smaller vessels, the arterioles, which lead to the capillaries. The capillaries are present in the form of microcirculation networks (capillary beds) in the organs and tissues. Exchange of metabolites and transport through the vessel wall is only possible in the capillaries, as only here the blood flow is sufficiently reduced (about 0.3mm/sec) and the vessel wall sufficiently thin.

On the return route to the heart the blood flows in venules, small veins and large veins. Arterial blood in the Systemic Circulation is richly oxygenated, whereas the venous blood has little oxygen. In the Pulmonary Circulation the arterial blood is poorly oxygenated, whereas the venous blood, are highly oxygenated (having replenished its oxygen supplies in the lungs).

Endothelial Cells

The endothelial cells are derived from embryonic mesenchyme and should not be regarded as epithelial, but as connective tissue cells. Endothelial cells line the lumina of all the vessels of the blood vascular and lymphatic vascular systems. Endothelial cells lining the blood vessels are very flattened, elongated cells, with elongated nuclei that protrude into the lumina. The total number of endothelial cells in the body is enormous (estimated as $6x10^{23}$ cells) and cover a very large surface area (700-1000m^2) and in total weigh about 1.5kg.

Blood Capillaries

Blood capillaries have a diameter of about 7-9mm, which is close to the dimensions of erythrocytes (about 7.2mm). The diameter of the capillaries varies according to the functional status of the tissue or organ. When functional demands rise the diameter of the capillaries enlarges, allowing increased exchange of oxygen and metabolites.

There are three different types of capillaries, however the differences are only visible at the ultrastructural level (by light microscopy these differences are not detectable):

- Continuous capillaries
- Fenestrated capillaries
- Sinusoids.

Continuous Capillaries

These have continuous endothelial cells located on a continuous basal lamina. In cross sections they are seen to be composed of 2-3 endothelial cells, connected by tight junctions. In the region of the contact between the ends of two endothelial cells there is a marginal fold, a small area in which the edge of one of the cells protrudes into the lumen. Continuous capillaries are characterized by abundant small invaginations of the cell surfaces (caveolae) and numerous micropinocytotic vesicles in the cytoplasm. All the materials crossing the cell (transcellular transport), in both directions do so via these micropinocytotic vesicles.

Continuous capillaries are found in those organs that need strict control on access of the substances from the blood. These include all the organs with a "blood-barrier" such as the "blood-brain-barrier" of the Central Nervous System or the "blood-thymus barrier".

Fenestrated Capillaries

These possess endothelial cells with groups of very small "pores" or "fenestrae", about 80-100nm diameter. These are seen in transmission electron micrographs and in particular after freeze-etching techniques. Fenestrated capillaries are common in most of the endocrine glands. One prominent site for fenestrated capillaries is in the renal glomeruli. The fenestrated capillaries lie on a continuous basal lamina.

Sinusoids

Sinusoids are irregular vessels with large diameters (30-40nm). In most cases the sinusoids are not cylindrical. Sinusoids are found in the liver, endocrine glands and in the hematopoietic organs (bone marrow, spleen). In many cases the sinusoids are also fenestrated. This is the case in those organs which need a very rich blood supply including most of the endocrine glands (hypophysis, suprarenal cortex, pancreas). Phagocytes are commonly associated with the walls of the sinusoids.

The exchange of materials through capillary walls can be:

transcellular via:

- micropinocytotic vesicles in the endothelium (as in continuous capillaries)
- fenestrations (as in fenestrated endothelium or sinusoids)

or intercellular via:

- gap junctions
- spaces between endothelial cells (as in sinusoids of spleen, liver).

Pericytes (Perivascular Cells)

Many capillaries have inconspicuous, elongated cells, similar in appearance to embryonic mesenchymal cells, associated with them. These cells, known as pericytes, or perivascular cells, are quite difficult to see in most histological preparations. These pericytes appear to have important roles in repair of blood vessels and connective tissue after injury. They have the potential to develop into fibroblasts, smooth musccells and may even be phagocytic.

Endothelial cells are known to produce a variety of local factors that are important in the functioning of the cardiovasystem. These include nitric oxide.

Morphology of Muscular (Distributing) Arteries

These vessels bring blood from the heart to the tissues in a high-pressure, fast-flow, system and consequently the arterial wall needs to be able to withstand the biomechanical stresses.

The arterial wall is composed of three main layers or tunics.

- *Tunica intima* (internal tunic) consisting of:
  - — endothelium (single lining layer of endothelial cells)
  - — sub-endothelial layer
  - — inner elastic limiting membrane (elastic lamina, which after fixation appears undulating).
- *Tunica media* (middle tunic) consisting of:
  - * circular smooth muscle (or spiral)
  - * concentric elastic lamina (formed by the smooth muscle cells).
- *Adventitia* (outer layer) composed of:
  - — connective tissue surrounding the vessel

— outer elastic limiting membrane (on the border between the *Tunica media* and the *Adventitia*
— *Vasa vasorum*. These are small blood vessels supplying oxygen and nutrients to the wall of the artery. The blood flow in the arterial lumen is too great for exchange of oxygen or nutrients.

Morphology of Elastic Arteries

These are arteries closest to the heart and need to withstand stresses of extremely high blood flow. Their structure is best seen in the aorta. They are called elastic arteries as their Tunica media possesses 50-75 well-developed elastic lamina in between the thick smooth muscle bands. These have a similar appearance to the inner elastic limiting membrane.

The elastic lamina prevent the excessive expansion of the vessel diameter and when they spring back they push the blood onwards i.e. they provide a shock-absorber effect permitting a more continuous blood flow despite the intermittent action of the heart.

The elastic lamina absorb the intermittent impact of the cardiac pulse. During diastole (inactive heart), the large elastic arteries return to normal size impelling the blood forward. This contributes to a more constant arterial pressure and blood flow. The function of the elastic arteries can be considered as an auxiliary pump, when no forward pressure is exerted by the heart.

Smooth Muscle in Arteries

The smooth muscle in arteries is important in maintaining the vessel diameter during blood flow and also plays a role in blood pressure levels. The vascular smooth muscle is in a state of tonus (partial contraction).

The degree of tonus of the smooth muscle cells in the wall of arteries and arterioles is controlled by the autonomic nervous system and also by endocrine secretions.

In hypertension (high blood pressure), often associated with stress or aging, the peripheral arterial vessels show increased tonus.

Arterioles

Arterioles are small vessels with a diameter of 0.5mm or less.

They consist of three basic layers:

— *Tunica intima* with endothelium alone (no subendothelial layer) and a very thin inner elastic limiting membrane
— *Tunica media* with only 4-5 layers of smooth muscle
— *Adventitia* that is fairly thin.

Metarterioles

These are small vessels that are on the border between arterioles and the capillary bed. They can act as sphincters and cut off the flow of blood into the capillary bed.

Arteriovenous Anastomoses

These represent direct connections between arterioles and venules. When there is no need for blood flow in the capillary bed these permit direct blood passage (arterial-venous-shunt). Arteriovenous anastomoses are very common in the dermis of the skin.

Anatomical and Functional End Arteries

Anatomical end arteries are vessels whose terminal branches do not anastomose. In the event that these vessels become blocked (atherosclerosis, blood clot) the tissues will be deprived of oxygen and an "infarct" develops.

Functional end arteries have anastomoses and in the event of blockage, alternative routes for blood and oxygen are available.

Arterial pathology has major clinical importance in medicine. Common arterial disorders include: atherosclerosis (fatty deposits and occlusion), arteriosclerosis (hardening of the arteries), hypertension, aneurysms (ballooning of the vessel). Cardiac infarct and cerebral infarct resulting from occlusion of the lumen of arteries are major causes of morbidity.

Veins

The veins constitute a low-pressure system of vessels. The return of blood to the heart from the capillary beds of the tissues follows a route of small venules, small veins and large veins. As the venous vessels near the heart they become large and with thicker walls. The route of the veins is in parallel to that of the arteries.

Characteristics of veins:

* more numerous than arteries
* diameter of vessels is larger than that of adjacent arteries
* walls of veins are thinner and less elastic or distensible than arteries. (As a result in histological preparations the lumen often appears collapsed or irregular)
* the relative numbers of *vasa vasorum* are greater in the veins (necessary as the vessels have much less oxygenated blood)
* valves are found in veins.

Veins are classified as large, medium or small veins.

Veins have three layers (*Tunica intima*, *Tunica media* and *Adventitia*), however the borders between these layers are much less distinct than in arteries. In veins the smooth muscles of the *T. media* are all circular muscles, grouped in bundles, whereas muscles present in the adventitia are longitudinal.

The movement of blood in veins is passive. The muscles play a role in tonus. Most of the muscles in veins are present in the adventitia and are longitudinal. Valves are present in veins, especially in those that transport blood against the force of gravity, such as in the legs. The valves are composed of folds of the *Tunica intima* (endothelium and connective tissue). The valves prevent the backflow of blood. Weakness in the walls of veins can result in varicose veins and improper closure of the valves.

Venules

These have a very small diameter (20-50mm). In total preparations such as from the mesentery, the venule wall is so thin that the erythrocytes are visible, whereas they are not seen in the adjacent and parallel arterioles.

Post-Capillary Venules

In lymphatic organs, such as the lymph nodes, the post-capillary venules have high or cuboidal endothelium. These specialized venules permit the recirculation of lymphocytes (especially T-lymphocytes) from the blood to the lymph.

Umbilical Blood Vessels

The umbilical blood vessels are atypical. The single vein is unusual in that it is very muscular. The paired umbilical arteries are also unusual. The *T. intima* is composed only of endothelium. The *T. media* is composed of two muscle layers: an inner longitudinal layer and an outer circular layer. Elastic fibres are present throughout the media.

Blood Portal Systems

In typical configurations an artery or arteriole carrying oxygenated blood enters the capillary bed, where there is exchange of oxygen and metabolites, and the vessel exiting the capillary bed is a venule or vein with deoxygenated blood. Portal systems describe situations where the blood vessel leaving the capillary bed is of the same category as the blood vessel entering the capillary bed. (vein-capillary bed-vein or artery-capillary bed-artery). In a venous portal system (such as in the liver) a vein (hepatic portal vein) enters the capillary bed and a vein (hepatic vein) exits the capillary bed. A similar portal system is found in the hypothalmus-hypophysis. An example of an arterial portal system is found in the renal cortex. Afferent arterioles break up into the capillary bed of the glomerular tufts of the renal corpuscle and the blood exits in efferent arterioles.

Lymph Vascular System

Functions of the Lymph Vascular System

- The lymph vessels return to the blood extracellular fluid from connective tissue spaces. This system ensures the return of water, electrolytes and plasma proteins to the blood.
- The lymph vascular system plays a kerole in homeostasis of the volume of extracellular fluid.
- The lymph vascular system also returns lymphocytes from the lymph nodes to the blood.
- The system also transports immunoglobulins (antibodies) from the lymphnodes to the blood.

Lymph Capillaries

Lymph is a fairly clear, transparent fluid that flows (passively) in small lymph capillaries. These are found in most organs close to blood capillaries (an exception is the CNS). The lymph capillaries begin as small blind-ending tubes.

- Typical lymph capillaries have a diameter of 10-50mm only. Lymph vessels have very thin walls. The wall of the lymph capillary is composed of a single layer of endothelium (about 0.3m m thick). Lymph vessels are hard to detect in histological preparations as the lumen tends to collapse.
- Lymph capillaries (unlike endothelial cells of blood vessels) lack a basal lamina.
- No pericytes or adventitial cells afound.
- They lack marginal folds (as seen in ultrastructure of blood capillaries).
- The lumen is usually free of cells (in comparison with blood capillaries where erythrocytes and other blood cells are common).

Movement of lymph in the lymph vessels is entirely passive. Valves, constructed from endothelial cells, prevent backflow of lymph. The lymph capillaries eventually drain into larger lymph vessels, with large lumina and thin walls and these ultimately drain into two large lymphatic ducts in the area at the base of the neck (thoracic duct, right lymphatic duct). These vessels return the lymph to the blood.

Circulatory System

The circulatory system is an organ system that passes nutrients (such as amino acids and electrolytes), gases, hormones, blood cells, nitrogen waste products, etc. to and from cells in the body to help fight diseases and help stabilize body temperature and pH to maintain homeostasis. This system may be seen strictly as a blood distribution network, but some consider the circulatory system as composed of the cardiovascular system, which distributes blood, and the lymphatic system, which distributes lymph. While humans, as well as other vertebrates, have a closed cardiovascular system (meaning that the blood never leaves the network of arteries, veins and capillaries), some invertebrate groups have an open cardiovascular system. The most primitive animal phyla lack circulatory system. The lymphatic system, on the other hand, is an open system.

The main components of the human circulatory system are the heart, the blood, and the blood vessels. The circulatory system includes: the pulmonary circulation, a "loop" through the lungs where blood is oxygenated; and the systemic circulation, a "loop" through the rest of the body to provide oxygenated blood. An average adult contains five to six quarts (roughly 4.7 to 5.7 liters) of blood, which consists of plasma, red blood cells, white blood cells, and platelets. Also, the digestive system works with the circulatory system to provide the nutrients the system needs to keep the heart pumping.

Two types of fluids move through the circulatory system: blood and lymph. The blood, heart, and blood vessels form the cardiovascular system. The lymph, lymph nodes, and lymph vessels form the lymphatic system. The cardiovascular system and the lymphatic system collectively make up the circulatory system.

Pulmonary Circulation

Pulmonary circulation is the portion of the cardiovascular system which transports oxygen-depleted blood away from the heart, to the lungs, and returns oxygenated blood back to the heart. Oxygen deprived blood from the vena cava enters the right atrium of the heart and flows through the tricuspid valve into the right ventricle where it is pumped through the pulmonary semilunar valve into the pulmonary arteries which go to the lungs. Pulmonary veins return the now oxygen-rich blood to the heart, where it enters the left atrium before flowing through the mitral valve into the left ventricle. Also, from the left ventricle the oxygen-rich blood is pumped out via the aorta, and on to the rest of the body.

Systemic Circulation

Systemic circulation is the portion of the cardiovascular system which transports oxygenated blood away from the heart, to the rest of the body, and returns oxygen-depleted blood back to the heart. Systemic circulation is, distance-wise, much longer than pulmonary circulation, transporting blood to every part of the body *except* the lungs.

Coronary Circulation

The coronary circulatory system provides a blood supply to the heart. As it provides oxygenated blood to the heart, it is by definition a part of the systemic circulatory system.

Heart

The heart pumps oxygenated blood to the body and deoxygenated blood to the lungs. In the human heart there is one atrium and one ventricle for each circulation, and with both a systemic and a pulmonary circulation there are four chambers in total: left atrium, left ventricle, right atrium and right ventricle. The right Atrium, which is the upper chamber of the right side. The blood that is returned to the right atrium is deoxygenated (poor in oxygen) and passed into the right ventricle to be pumped through the pulmonary artery to the lungs for re-oxygenation and removal of carbon dioxide. The left atrium receives newly oxygenated blood from the lungs as well as the pulmonary vein which is passed into the strong left ventricle to be pumped through the aorta to the tissues of the body.

Closed Cardiovascular System

The cardiovascular systems of humans are closed, meaning that the blood never leaves the system. In contrast, oxygen and nutrients diffuse across the blood vessel layers and enters interstitial fluid, which carries oxygen and nutrients to the target cells, and carbon dioxide and wastes in the opposite direction. The other component of the circulatory system, the lymphatic system, is not closed.

Other Vertebrates

The circulatory systems of all vertebrates, as well as of annelids (for example, earthworms) and cephalopods are *closed*, just as in humans. Still, the systems of fish, amphibians, reptiles, and birds show various stages of the evolution of the circulatory system. In fish, the system has only one circuit, with the blood being pumped through the capillaries of the gills and on to the capillaries of the body tissues. This is known as *single cycle* circulation. The heart of fish is therefore only a single pump (consisting of two chambers). In amphibians and most reptiles, a double circulatory system is used, but the heart is not always completely separated into two pumps. Amphibians have a three-chambered heart.

In reptiles, the ventricular septum of the heart is incomplete and the pulmonary artery is equipped with a sphincter muscle. This allows a second possible route of blood flow. Instead of blood flowing through the pulmonary artery to the lungs, the sphincter may be contracted to divert this blood flow through the incomplete ventricular septum into the left ventricle and out through the aorta. This means the blood flows from the capillaries to the heart and back to the capillaries instead of to the lungs. This process is useful to ectothermic (cold-blooded) animals in the regulation of their body temperature.

Birds and mammals show complete separation of the heart into two pumps, for a total of four heart chambers; it is thought that the four-chambered heart of birds evolved independently from that of mammals.

Open Circulatory System

The Open Circulatory System is a system in which fluid (called hemolymph) in a cavity called the hemocoel bathes the organs directly with oxygen and nutrients and there is no distinction between blood and interstitial fluid; this combined fluid is called hemolymph or haemolymph. Muscular movements by the animal during locomotion can facilitate hemolymph movement, but diverting flow from one area to another is limited. When the heart relaxes, blood is drawn back toward the heart through open-ended pores. Hemolymph fills all of the interior hemocoel of the body and surrounds all cells. Hemolymph is composed of water, inorganic salts (mostly Na^+, Cl^-, K^+, Mg^{2+}, and Ca^{2+}), and organic compounds (mostly carbohydrates, proteins, and lipids). The primary oxygen transporter molecule is hemocyanin.

There are free-floating cells, the hemocytes, within the hemolymph. They play a role in the arthropod immune system.

Absence of Circulatory System

Circulatory systems are absent in some animals, including flatworms (phylum Platyhelminthes). Their body cavity has no lining or enclosed fluid. Instead a muscular pharynx leads to an extensively branched digestive system that facilitates direct diffusion of nutrients to all cells. The flatworm's dorso-ventrally flattened body shape also restricts the distance of any cell from the digestive system or the exterior of the organism. Oxygen can diffuse from the surrounding water into the cells, and carbon dioxide can diffuse out. Consequently every cell is able to obtain nutrients, water and oxygen without the need of a transport system.

Some animals, such as jellies, have more extensive branching from their gastrovascular cavity (which functions as both a place of digestion and a form of circulation), this branching allows for bodily fluids to reach the outter layers, since the digestion begins in the inner layers.

Measurement Techniques

- Electrocardiogram — for cardiac electrophysiology
- Sphygmomanometer and stethoscope — for blood pressure
- Pulse meter — for cardiac function (heart rate, rhythm, dropped beats)
- Pulse — commonly used to determine the heart rate in absence of certain cardiac pathologies
- Heart rate variability — used to measure variations of time intervals between heart beats
- Nail bed blanching test — test for perfusion
- Vessel cannula or catheter pressure measurement — pulmonary wedge pressure or in older animal experiments.

Health and Disease

Oxygen Transportation

About 98.5% of the oxygen in a sample of arterial blood in a healthy human breathing air at sea-level pressure is chemically combined with haemoglobin molecules. About 1.5% is physically dissolved in the other blood liquids and not connected to Hgb. The haemoglobin molecule is the primary transporter of oxygen in mammals and many other species.

History of Discovery

The earliest known writings on the circulatory system are found in the Ebers Papyrus, an ancient Egyptian medical papyrus containing over 700 prescriptions and remedies, both physical and spiritual. In the papyrus, it acknowledges the connection of the heart to the arteries.

The Egyptians thought air came in through the mouth and into the lungs and heart. From the heart, the air travelled to every member through the arteries. Although this concept of the circulatory system is greatly flawed, it represents one of the earliest accounts of scientific thought.

The knowledge of circulation of vital fluids through the body was known to Sushruta. He also seems to have possessed knowledge of the arteries, described as 'channels' by Dwivedi & Dwivedi (2007). The valves of the heart were discovered by a physician of the Hippocratean school around the 4th century BC. However their function was not properly understood then. Because blood pools in the veins after death, arteries look empty.

Ancient anatomists assumed they were filled with air and that they were for transport of air. Greek physician Herophilus distinguished veins from arteries but thought that the pulse was a property of arteries themselves. Greek anatomist Erasistratus observed that arteries that were cut during life bleed. He ascribed the fact to the phenomenon that air escaping from an artery is replaced with blood that entered by very small vessels between veins and arteries. Thus he apparently postulated capillaries but with reversed flow of blood.

The 2nd century AD, Greek physician, Galen, knew that blood vessels carried blood and identified venous (dark red) and arterial (brighter and thinner) blood, each with distinct and separate functions. Growth and energy were derived from venous blood created in the liver from chyle, while arterial blood gave vitality by containing pneuma (air) and originated in the heart. Blood flowed from both creating organs to all parts of the body where it was consumed and there was no return of blood to the heart or liver. The heart did not pump blood around, the heart's motion sucked blood in during diastole and the blood moved by the pulsation of the arteries themselves.

Galen believed that the arterial blood was created by venous blood passing from the left ventricle to the right by passing through 'pores' in the interventricular septum, air passed from the lungs via the pulmonary artery to the left side of the heart. As the arterial blood was created 'sooty' vapors were created and passed to the lungs also via the pulmonary artery to be exhaled.

In 1242, the Arabian physician, Ibn al-Nafis, became the first person to accurately describe the process of pulmonary circulation, for which he is sometimes considered the father of circulatory physiology. Ibn al-Nafis stated in his *Commentary on Anatomy in Avicenna's Canon*:

> *"...the blood from the right chamber of the heart must arrive at the left chamber but there is no direct pathway between them. The thick septum of the heart is not perforated and does not have visible pores as some people thought or invisible pores as Galen thought. The blood from the right chamber must flow through the vena arteriosa to the lungs, spread through its substances, be mingled there with air, pass through the arteria venosa to reach the left chamber of the heart and there form the vital spirit..."*

Finally William Harvey, a pupil of Hieronymus Fabricius (who had earlier described the valves of the veins without recognizing their function), performed a sequence of experiments, and published *Exercitatio Anatomica de Motu Cordis et Sanguinis in Animalibus* in 1628, which "demonstrated that there had to be a direct connection between the venous and arterial systems throughout the body, and not just the lungs.

Most importantly, he argued that the beat of the heart produced a continuous circulation of blood through minute connections at the extremities of the body. This is is a conceptual leap that was quite different from Ibn al-Nafis' refinement of the anatomy and bloodflow in the heart and lungs." This work, with its essentially correct exposition, slowly convinced the medical world. However, Harvey was not able to identify the capillary system connecting arteries and veins; these were later described by Marcello Malpighi.

Circulatory System

The Circulatory system is an organ system that helps maintain homeostasis. It delivers gases, helps with the regulation of temperature, and the transport of nutrition and waste. It is also known as the cardiovascular system.

It consists of heart, blood, and blood vessels. There are two different types of the circulatory system: open circulatory system and closed circulatory system. Most of vertebrates have a closed circulatory system, whereas most invertebrate have an open circulatory system.

Vascular Pathway

Human Vascular Pathway

Pulmonary Circuit: The Pulmonary circuit is involved in the exchange of gases in the lungs. Blood flows from the right ventricle of the heart into the pulmonary artery. From here, it branches both left and right directions, making it's way toward the lungs. Once it reaches the lungs, carbon dioxide is released and oxygen is taken in, both in relation to their concentration gradient. In this case, some carbon dioxide remains in the blood and some oxygen remains in the lungs. However, under hyperbolic conditions, such as in a hyperbolic oxygen chamber, the level and pressure of the oxygen is so high that a much larger amount of oxygen is consumed by the blood, forcing a larger quantity of carbon dioxide out of the blood.

The oxygen level inside the blood and the oxygen level inside the lungs will attempt to balance out. When the oxygen level is seriously depleted in an environment, the body will produce more hemoglobin to decrease the oxygen:blood ratio and to provide more carriers for the oxygen molecules. In hyperbolic conditions, the oxygen without hemoglobin quantity will increase, while the hemoglobins will also fill to capacity.

The blood then leaves the lung capillaries and travels down the pulmonary vein and enters the left atrium of the heart.

Systemic Circuit

The systemic circuit is involved in carrying oxygen rich blood to the body tissues and organs. The blood leaves the left ventricle and enters the right ventricle.

The AV node of the heart signals ventricular contraction and the blood is moved through the aortic semilunar valve, into the aorta. The aorta is the largest and strongest artery in the body, branching into other arteries which spread throughout the body. The blood then returns through venules and veins to the superior and inferior vena cava before entering the right atrium.

Bibliography

A P V Samvel: *Agri-Business Management*, Satish Serial Pub, Delhi, 2008.

A.J. Macself: *Soils and Fertilizers*, Satish Serial Pub, Delhi, 2005.

Akhil Baruah: *Advanced Morphology of Angiosperms*, Aavishkar, Delhi, 2008.

Alfred Steferud: *Diseases of Fruits and Nuts*, Biotech Books, Delhi, 2005.

Andrews L.: *Citrus Production - Orange*, St. Augustine, Trinidad and Tobago, 1990

Berger, P.L.: *Pyramids of Sacrifice: Political Ethics and Social Change*, New York, Basic Books, 1974.

C L Goswami: *Plant Physiology : Research Methods*, Scientific, Delhi, 2007.

C.R. Sudharmai Devi: *Analytical Procedures in Soil Science and Agricultural Chemistry*, Agrotech, Delhi, 2004.

Chambers, R.: *Rural Development: Putting the Last First*, London, Longman, 1983.

Clawson, Calvin: *The Mathematical Traveller*, New York, Plenum Press, 1994.

Crump, Thomas: *The Anthropology of Numbers*, Cambridge, Cambridge University Press, 1990.

D.H. Robinson; *Entomology : Principles and Practices*, Agrobios, Delhi, 2001.

Daniel Dean: *Growing and Breeding of Potatoes*, Asiatic Pub, Delhi, 2006.

David White: *The Physiology and Biochemistry of Prokaryotes*, Oxford University Press, Delhi, 2007.

Dharamvir Hota: *Modern Biotechnology in Plant Breeding*, Gene Tech Books, Delhi, 2007.

Fortuner, R.: *Nematode Identification and Expert System Technology*, New York, Plenum Press, 1988.

G K Ghosh: *Biopesticide and Integrated Pest Management*, APH, Delhi, 2009.

G.N.V. Febulaus: *Embryology and Apomixis in Grasses*, Regency, Delhi, 2000.

Gilbert Wooding Robinson: *Soils : Their Origin Constitution and Classification*, Biotech Books, Delhi, 2005.

Govind Prasad: *Environmental Geomorphology*, Discovery Publishing House, Delhi, 2008.

Govind Prasad: *Trends and Techniques of Geomorphology*, Discovery, Delhi, 2007.

Hartley W.: *A Checklist of Economic Plants in Australia*, Melbourne, C.S.I.R.O., 1979.

I S Pawar: *Trends in Wheat Breeding*, CBS Pub, Delhi, 2006.

K M Sellamuthu: *Fundamentals of Forest Soils*, Satish Serial Pub, Delhi, 2008.

Keshav Prasad Yadav: *Application of Morphometry in Geomorphology*, Radha Pub, Delhi, 2008.

Madan Lal Bagdi: *Physiology, Biochemistry and Biotechnology*, Manglam Pub, Delhi, 2007.

Mitra S.: *Postharvest Physiology and Storage of Tropical and Subtropical Fruits*, Oxon, CABI, 1997.

P.R. Yadav: *Textbook of Embryology*, Campus Books, Delhi, 2001.

Proctor F. J.: *Onions in Tropical Regions*, Kent, Natural Resources Institute, 1990.

R K Yadav: *World Fishes and Aquaculture*, Navyug, Delhi, 2008.

R. Mishra: *Morphology of Fungi*, Discovery Pub, Delhi, 2005.

R.C. Sharma: *Diseases of Horticultural Crops: Fruits*, Indus, Delhi, 1999.

Rajni Sharma: *An Introduction to Plant Morphology*, Campus, Delhi, 2004.

Renuka Desai: *Plant Physiology*, Adhyayan, Delhi, 2008.

Samiullah Khan: *Plant Breeding Advances and in vitro Culture*, CBS, Delhi, 1997.

Sanchez, P. A.: *Properties and Management of Soils in the Tropics*, John Wiley and Sons, New York, 1976.

T V Sathe: *Biotechnological Approaches in Entomology*, Manglam Pub, Delhi, 2008.

U K Mishra: *Physiology of Animal Growth and Bioenergetics*, Satish Serial Pub House, Delhi, 2008.

V C Baligar and R B Clark: *Physiology of Crop Production*, International Book, Delhi, 2007.

Vardhan Bhaskar: *Animal Physiology*, Campus Books, Delhi, 2008.

Vikas Chaudhary: *Entomology and Pest Management*, Navyug, Delhi, 2008.

W L H Duckworth: *Morphology and Anthropology : A Handbook for Students*, Cosmo, Delhi, 2006.

White, G.F.: *Natural Hazards: Local, National, Global*, Oxford University Press, New York, 1974.

Index

□□□